Repairing Infrastructures

Infrastructures Series

edited by Geoffrey C. Bowker and Paul N. Edwards

Repairing Infrastructures

The Maintenance of Materiality and Power

Christopher R. Henke and Benjamin Sims

The MIT Press
Cambridge, Massachusetts
London, England

The open access edition of this book was made possible by generous funding from Arcadia—a charitable fund of Lisbet Rausing and Peter Baldwin.

ARCADIA
A charitable fund of Lisbet Rausing and Peter Baldwin

This book was set in ITC Stone Serif Std and ITC Stone Sans Std by New Best-set Typesetters Ltd.

Library of Congress Cataloging-in-Publication Data

Names: Henke, Christopher, 1969- author. | Sims, B. (Benjamin), author.
Title: Repairing infrastructures : the maintenance of materiality and power / Christopher R. Henke and Benjamin Sims.
Description: Cambridge, Massachusetts : The MIT Press, 2020. | Series: Infrastructures series | Includes bibliographical references and index.
Identifiers: LCCN 2020002135 | ISBN 9780262539708 (paperback)
Subjects: LCSH: Public works—United States. | Infrastructure (Economics)— United States.
Classification: LCC HD3885 .H44 2020 | DDC 363.6068—dc23
LC record available at https://lccn.loc.gov/2020002135

149115289

For Carolyn and Kathy

Contents

Acknowledgments

We thank the many colleagues and friends who supported this project and helped us see it into print.

We first met during our graduate studies in the Science Studies Program at the University of California, San Diego. This book reflects the influence of both San Diego itself, through case studies in chapters 2 and 3, and the intellectual community at UCSD, which shaped our thinking about repair in important ways. Special thanks are due to Chandra Mukerji, an important mentor to both of us. This book owes a crucial intellectual debt to her work on infrastructures and material culture, and has been shaped in many other ways by methods and perspectives she introduced us to. Thank you, Chandra, for your continued support and inspiration over the years. We are also indebted to many other colleagues from UCSD, including Patrick Carroll, Mark Jones, Martha Lampland, Charlie Thorpe, Josh Dunsby, Jennifer Jordan, Steven Shapin, and other colleagues and mentors we first met as graduate students, including Trevor Pinch, Tom Gieryn, Mike Lynch, and Harry Collins.

We are fortunate to be part of a growing community of scholars working on repair, and we have benefited from the insights and friendly critiques of many colleagues, especially participants in the International Comparative Urban Retrofit workshop in 2012, the Repair Work Ethnographies workshops in 2014 and 2015, and the workshop on Maintenance, Repair and Beyond held in 2015. We are grateful to have been included in these conversations as well as the resulting publications: *Retrofitting Cities: Priorities, Governance and Experimentation* (Mike Hodson and Simon Marvin, editors; Routledge, 2016); *Repair Work Ethnographies: Revisiting Breakdown, Relocating Materiality* (Ignaz Strebel, Alain Bovet, and Philippe Sormani, editors;

Palgrave Macmillan, 2019); and an issue on repair in the online journal *Continent* (Lara Houston, Daniela K. Rosner, Steven J. Jackson, and Jamie Allen, editors; issue 6.1, 2017). The editors and peer reviewers for each of these projects provided helpful feedback that has shaped our thoughts for this book.

Colleagues who heard presentations on portions of this project at UCSD's Science Studies Program, Cornell University's Department of Science and Technology Studies, and conference sessions for the Society for Social Studies of Science and the American Sociological Association also provided valuable feedback.

This work would not have been possible without the patience and trust of the people we interviewed, observed, and just chatted with in the course of developing the case studies in this book. These included the maintenance technicians described in chapter 2, the Caltrans and UCSD engineers and other specialists we discuss in chapter 3, and the nuclear weapons scientists and engineers featured in chapter 4. Chapter 3 would not exist without the amazing, largely uncompensated work of the activists and muralists who created Chicano Park in San Diego's Barrio Logan and have maintained it as an artistic landmark and community resource for over forty years. We owe particular thanks to Salvador Torres, who shared many insights on the history of the park and murals with us, and Mario Torero, who graciously allowed us to use an image of his mural *Colossus* on the cover of this book.

We thank Infrastructures series coeditor Paul Edwards for encouraging us to write this book as part of the series, an idea that originated during an unexpected, fortuitous meeting between Ben and Paul in Los Alamos on a wintry day in 2013. Series coeditor Geoff Bowker and MIT Press editors Justin Kehoe and Katie Helke also provided important feedback and helped us through the review process. Thanks also to several anonymous reviewers for their detailed comments on our manuscript, editors Ginny Crossman, Bev Miller, and Cadi Klepeis for their support in the final production of the book, and Amron Gravett of Wild Clover Book Services for her work on the book's index. We gratefully acknowledge grant support from Colgate University's Research Council for book production costs, and Arcadia's support of open access publishing.

Ben's thanks. Thanks to Chris for giving me the opportunity to work within and expand on a universe of repair concepts that originates in his work. Our continued engagement has been a lifeline to the world of

academic sociology and science and technology studies (STS) while working at an institution where I have few colleagues in these fields. Thanks for
being a great friend and collaborator.

I am grateful to many friends and colleagues at Los Alamos National
Laboratory who have helped me take my background in STS and sociology
in directions I never thought possible, and taught me a great deal about science, technology, and infrastructure along the way. They include members
of the former systems ethnography team, especially Greg Wilson, Laura
McNamara, and Andrew Wiedlea, and others including John Ambrosiano,
Kari Sentz, Joanne Wendelberger, Linn Collins, Walt Gilmore, Trinity Over-
myer, and Bob Benjamin. And special thanks to my colleagues in the statis-
tical sciences group at Los Alamos who, despite the fact that I have never
taken a single statistics class, have provided me with an amazingly supportive organizational home for over fifteen years. I appreciate the willingness
of my managers to allow me the flexibility and time off to complete this
book as an independent project. Of course, the standard disclaimer applies:
the views expressed in this book are not necessarily shared or endorsed by
any institution or person mentioned here.

Finally, I thank my family for their love and support, including my parents, Chris and Cathie; my children, Sophie and Nathan; and my wife,
Kathy, who inspires me to pay attention to the important stuff in all aspects
of life.

Chris's thanks. My first thanks are to Ben for being a friend and collaborator for nearly thirty years. Conversations and shared work together,
including this book, have had a profound impact on my thinking about
repair and related topics. I especially appreciate the deep way that he works
through and questions concepts, which has been a good check on my tendency to make things up as I go. Thanks for sticking with me and helping
me to become a better writer and thinker.

I owe much to my colleagues at Colgate University, especially those who
have read and commented on my work over two decades. I have learned so
much from them, including wisdom about how to teach, write, and serve.
Elana Shever, John Pumilio, Michelle Bigenho, Jon Hyslop, Santiago Juarez,
Paul Lopes, Andy Pattison, Nancy Ries, Chandra Russo, and Alicia Simmons
read and commented on early portions of this project. Many of the topics
in this book were shaped and inspired by conversations and shared work
with my students in the context of courses and research partnerships. I

am also blessed to have a great group of friends and colleagues who keep life fun and meaningful, especially members of Hamilton Bible Fellowship, Your Elders and Betters, and the Hamilton Climate Preparedness Working Group.

Special thanks to my running buddies, Mark Stern and Rob Nemes, for all the miles and many fun and random conversations. Thanks to them also for their comments on portions of this manuscript and for listening to me talk about repair as we jog past the spillway and our other well-traveled routes in central New York State.

Final thanks are due to my family, including my parents, Doris and Russ; my daughter, Lin; and the best wife and fellow-life-traveler I could ever hope for, Carolyn.

1 Introduction: A Tool Kit for Understanding Infrastructure and Repair

Bridge Collapse in Minneapolis

On Wednesday, August 1, 2007, at 6:05 p.m., the Interstate 35W bridge over the Mississippi River near downtown Minneapolis suddenly and completely collapsed without warning, sending a roadway full of rush-hour traffic plunging more than one hundred feet into the river below.[1] Surveillance video shows the steel truss supporting the central span of the roadway giving way and dropping into the river in a cloud of spray, followed by the rapid collapse of the rest of the bridge.[2] Some approaching drivers barely succeeded in stopping as the road in front of them vanished. Others were unable to stop and drove over the edge. Altogether, 111 vehicles were caught up in the collapse, including a school bus full of children; 145 people were injured and 17 died. Fortunately, quick action by bystanders and emergency responders led to the rescue of almost everyone who survived the initial collapse, including all the students on the bus.[3]

An event like this is shocking not just because of its human toll, but also in the way it upends expectations. Infrastructures like roads, sewers, and mobile phone networks are so central to daily life that when they are reliable, they tend to fade from conscious awareness. Even an iconic landmark like the Golden Gate Bridge, which is frequently visited, photographed, and seen in movies, is rarely appreciated for functional characteristics like the integrity of its girders, the stability of its footings, and its margin of safety under heavy traffic loads. The mundane, functional aspects of infrastructure tend to come to the surface when something has gone wrong: the reliability of the electrical grid might become a major topic of interest after a power outage; the locations of mobile phone towers might come to mind when cellular signal is lost. And while material objects like bridges or mobile phone towers are easy enough to track down if you choose to look for them, there is another mundane aspect of infrastructure that is more ephemeral: the constant repair and maintenance work crucial to keeping any infrastructure in working order decades after it is built. This work might not be noticeable on a daily basis, but when a bridge collapses suddenly after more than forty years of unproblematic service, it is natural to start thinking about whether it was in good repair.

So it went with the I-35W bridge collapse: a bridge that few people, even those who drove over it every day, paid much attention to suddenly became a top story on the national news, with a focus on its repair and maintenance.[4] Of particular interest was the fact that inspectors had rated the condition of the bridge superstructure as "poor" for seventeen consecutive years prior to the collapse, leading to a federal designation of the bridge as "structurally deficient."[5] Reporters connected this to a broader narrative of nationwide infrastructure decay (which we discuss in more detail later in this chapter), a rare moment of focused national attention on infrastructure repair and maintenance and an opportunity for the general public to ponder some uncomfortable questions, such as: Is the infrastructure beneath our feet really as strong and solid as it seems? Are we as a nation placing enough of a priority on infrastructure maintenance? Who does all that repair and maintenance work anyway, and are they doing a good job? And what prevents bridges that we drive over every day from collapsing underneath us? Of course, as the disaster inevitably faded from memory, people went right back to driving over bridges without asking themselves these questions. More than a decade later, we would guess that most readers of this book have only a hazy recollection of this event, if they heard of it at all. Our goal for this book is to keep the focus on these uncomfortable questions and, more importantly, provide a set of tools and concepts to make sense of how repair and maintenance keep infrastructure in working order.

Repair as Social and Technical Work

Repair, as we use the term in this book, is the work required to maintain technologies of all kinds—from heroic efforts in moments of breakdown and crisis to the mundane and hidden maintenance work that keeps things running day-to-day.[6] The main aims of this book are to show how the enduring function and influence of infrastructure are made possible by the constant work of repair and to explore the causes and consequences of the strange, ambivalent, and increasingly central role of infrastructure repair in our lives today. To address these issues, we take a broad view of repair, which we describe briefly before returning to the I-35W bridge example.

While today the term *repair* is perhaps most often used to describe the hands-on work of restoring material things to working order, it has been used to describe the act of restoring or rebuilding all manner of things that are "damaged, worn, or faulty." These include objects and structures; cities; countries; a person's appearance or health; bodies and body parts; and immaterial things like friendships, honor, status, and finances.[7] In a more specialized use in sociology, *repair* is also used to refer to the methods we use to overcome misunderstandings and interruptions in conversations.[8]

We embrace all of these uses of *repair* because we see this process as fundamentally one of restoring both social and material order—a crucial point. Although we focus on material repair in this book, we argue that it almost always goes along with repairs to other forms of order. So, for example, restoring a fallen bridge may also serve to repair a city's transportation network and public trust in engineers and politicians; repairing an office air-conditioning vent may also involve gently adjusting an office worker's expectations about temperature and airflow; and retrofitting a bridge to resist earthquakes may also involve rebuilding trust between a state transportation agency and community activists, as well as maintaining the appearance of murals painted on the bridge columns. (These are all case studies we return to later in the book.)

More specifically, our perspective on repair emphasizes (1) the broad range of activities that go on under the heading of repair; (2) the dual social and technical nature of most material repair work; and (3) the variety of scales of repair work, from local fixes to broad, systemic efforts. We address each of these in turn.

Our first key point about repair emphasizes its ubiquity: we all engage in repair work on a daily basis, even when we might not realize we are doing it. Everyday aspects of repair work include hitting a machine just the right way so it stops making an annoying sound, debating with a family member about whether to continue repairing an older car that has seen better days, calling a tech support hotline to troubleshoot a computer problem, or emailing a local official to complain about the condition of roads in your neighborhood. In these cases, repair work is not always about directly fixing a piece of technology, and the actions are less specialized than those performed by an auto mechanic, tailor, or building contractor. Many of the examples we present in this book do focus on specialized repair workers, but in these cases, we show how their work is associated with broader discussions and arguments about what needs to be repaired, how it should be repaired, and even whether it is actually broken in the first place. These interactions can bring together a variety of individuals from different walks of life and play an important role in negotiating the evolving form and meaning of our infrastructures.

Expanding on our second point, looking at repair from a broad social and technological perspective provides a richer picture of infrastructures and how they function. Many infrastructures have a hard material structure, such as the columns and spans used to construct a bridge. By tracing the repair work of diverse actors, however, we start to see infrastructures as complex hybrids of material as well as social and political elements, and it is usually hard to separate them.

For this reason, we follow the lead of other scholars of science, technology, and society in describing infrastructures as *sociotechnical systems*.[9] This term emphasizes the tangled messiness of infrastructures, which are made not only of concrete, steel, and wires but also budget appropriations, engineering standards, and backroom dealmaking. A sociotechnical lens helps us see how repair can engage more diverse sets of people and things than we might expect, because the work of repair can include tinkering with any of these elements. In addition, understanding infrastructures as complex sociotechnical systems, with histories of contingency and change, makes it clear that they are part of our culture and politics. They are therefore connected to the broader structures of privilege, inequality, and justice that shape who has control and whose interests are ignored when it comes to building and repairing infrastructures. An infrastructure that seems to be working just fine to one group of people may seem in desperate need of repair to another.

Finally, addressing our third point, repair takes place at many different scales, particularly where infrastructural systems are concerned. Much of the hands-on work of infrastructure repair is done by individuals or small groups focused on solving local problems. But infrastructures can also fail in global ways, leading to the intervention of system operators, engineers, managers, and even political leaders, which can bring various interest groups and the general public into the process as well. On a global scale, international standard-setting organizations and negotiations among nation-states may also play an important role. Our view of repair encompasses the activities of all of these actors and their interactions: whether repairing infrastructures through logistical or political means, or through direct material engagement, a diverse range of players each have their own crucial roles in identifying and repairing infrastructure breakdowns.

Sociotechnical Repair and the I-35W Bridge

The I-35W bridge collapse is a useful example for understanding how some of these aspects of repair and maintenance come into play in a real-world infrastructural context. It is particularly relevant because much of the subsequent investigation centered around issues of repair and maintenance, meaning that we have an unusually deep record of the role of these activities in the life of this bridge. As we will see, the fact that this bridge collapsed does not mean there was anything particularly special about its repair or maintenance history; what is special is that we have such a good public record of these routine activities due to the investigation.

After opening for traffic in 1967, the bridge was inspected annually starting in 1971, as required by federal regulations.[10] This was a fairly involved process, taking five or six days, in which a team of inspectors closely examined every piece of the bridge—in some years from a required distance of 24 inches or less.[11] Inspection was mostly visual, with some help from tools for detecting cracks not yet visible.[12] These kinds of inspections are how bridge engineers identified repair and maintenance needs. Inspectors rated the bridge "structurally deficient" starting in 1991, mostly due to the rusty condition of its bearing devices.[13] Although this sounds serious, it did not indicate a major safety concern, but rather that the bridge was in a condition to be eligible for additional federal maintenance funding.[14] If inspectors felt there was a safety issue, they had the option of finding a "critical deficiency" and closing the bridge to traffic. They never took that step.

Beyond efforts to address rust and other more expected wear and tear, repair workers commonly fixed cracks in the steel, which was usually done by drilling out a hole 1.5 to 2.0 inches in diameter at each end of the crack, preventing it from spreading farther.[15] Occasionally workers would add reinforcements to the steel in areas where cracks had become particularly bad.[16] Eventually engineers realized the cracks were due to metal fatigue,[17] and the Minnesota Department of Transportation (MnDOT) funded several studies on ways of retrofitting the bridge structure to minimize this issue,[18] which were still underway when the bridge collapsed.[19]

Besides repairing wear and tear, MnDOT added several new features to the bridge over the years to address changing needs. First, in 1977, they increased the thickness of the concrete on the bridge deck to protect against corrosion due to road deicing chemicals. By 1998, the median barrier and railings on top of the bridge no longer met current safety standards, so these were removed and replaced. At the same time, a computer-controlled anti-icing system was added to address the large number of winter accidents on the bridge.[20] In 2007, contractors were in the process of grinding down and replacing the aging concrete of the bridge deck when the bridge collapsed.

Despite the prominence of repair and maintenance issues in the investigation, the National Transportation Safety Board (NTSB) ultimately concluded they had almost nothing to do with the collapse of the bridge. They instead found that the bridge was done in by a mistake in the design of its gusset plates, the metal plates that are used to attach bridge girders to one another: the plates were much too thin, and they would have eventually led to the collapse of the bridge even if it was in perfect condition.[21] This issue was missed throughout the lifetime of the bridge because gusset plates were generally built so much stronger than necessary that bridge engineers and inspectors never gave them much thought. So firm was this bias that even when inspectors had noticed that the gusset plates were bowed, they assumed it to be a relatively harmless construction defect, not a sign of imminent structural failure.[22]

Others vehemently disagreed with the NTSB's conclusions. In particular, construction attorney Barry LePatner has argued that the NTSB report downplays deterioration and maintenance problems with the bridge. He suggests that if efforts to retrofit the bridge to address the fatigue cracking issue and add structural redundancy had proceeded with more urgency, the resulting improvements might have prevented the collapse, and the gusset plate design error might have been noticed earlier.[23] And even the NTSB agreed that repairs to the bridge deck were, ironically, the immediate cause of the collapse: the weight of piles of construction materials, which ordinarily would not have been a problem, turned out to be the last straw that caused the weak gusset plates to give way.

After the bridge collapsed, additional repair concerns emerged around construction of a new bridge and how to prevent similar accidents in the future. For the Minnesota state legislature, the solution was to spend money to fix older bridges regardless of the NTSB findings. A few months after the collapse, the legislature passed a gas tax increase with bipartisan support that was used to fund a major bridge improvement program, with a focus on updating fracture-critical and structurally deficient bridges. They did this over the veto of Governor Tim Pawlenty, who used the NTSB findings to argue that the spending was unnecessary. Thanks to these improvements, the percentage of urban freeway traffic traveling over "structurally deficient" bridges in Minnesota had dropped from almost 6 percent to a little over 1 percent by 2016.[24]

For the NTSB, since their investigation identified a design error as the cause, the question became how to catch similar errors in the future. Instead of material changes to bridges, their recommendations called for improvements in design and inspection processes. These included new design quality assurance procedures, new bridge inspection procedures related to gusset plates, new procedures for assessing bridge load capacity, and new guidelines for placement of construction material.[25] Essentially the board identified breakdowns in the behavior of bridge engineers and inspectors, and the way to repair these breakdowns was to change the regulations that governed their actions. Again, this reveals that what keeps a bridge functioning is not just steel beams, gusset plates, rivets, and concrete, but an entire sociotechnical system that includes things like documents and training, professional standards, and organizational procedures.

The final question was what to do about the gaping hole that now existed in the center of the Minneapolis road transportation network. This is yet another level of repair, related to issues of scale. There was no longer any question of repairing the bridge itself, which lay in a tangled mess on the river banks. Instead, attention shifted to repairing the larger infrastructure system it was part of. In the short term, this meant rerouting traffic and adding travel lanes to alternative roadways.[26] In the longer term, it required replacing the bridge. With quick action to secure federal funds and an accelerated contracting and construction process, a state-of-the-art replacement bridge was completed and opened to traffic just thirteen months after its predecessor's collapse—quite a contrast to the drawn-out efforts to repair structural issues with the older bridge before its collapse.[27]

In the end, this story suggests two distinct ways in which maintenance and repair were integral to the life, and demise, of the bridge. First, repair and maintenance were absolutely essential to protect the bridge from degradation through aging and wear and tear. In concluding that inadequate repair and maintenance were *not* to blame for the collapse of the bridge, the NTSB's conclusions give short shrift to the role they may have played in preventing a collapse. Were it not for the extensive efforts of engineers to address fatigue cracking and other mundane impacts of aging, it seems quite possible that the bridge could have collapsed sooner than it did. Yet other efforts to prevent degradation, like adding concrete to the bridge deck to resist corrosion, may have indirectly contributed to the collapse by increasing the load on the structure—and, in the case of the final effort to refurbish the bridge deck, directly contributed to the collapse. The complex role of repair and maintenance in these events, however, makes it clear how essential these activities are to understanding the existence of infrastructure over time.

Second, the story of the I-35W bridge shows that not all repair and maintenance work is necessarily driven by material changes in infrastructure. The drawn-out efforts to address the fracture-critical design of the bridge are a case in point. These were driven not by aging or wear and tear, but by changes in engineering knowledge and safety standards over time, which led to features of the bridge that had seemed perfectly acceptable in the 1960s being seen as dangerous flaws only a few decades later. Through these developments, the materiality of the bridge was interpreted in new ways, leading it to appear very different to engineers by the 1990s. These changes in knowledge and standards are common during the life spans of major infrastructures and are another important driver of repair and maintenance work—which in this context are often described in terms of *retrofit*, a topic we explore in more detail in chapter 3.

In summary, the story of the I-35W bridge collapse illustrates the complex, materially and socially entangled nature of breakdown and repair, particularly with regard to infrastructure systems. The cast of characters includes bridge inspectors who did not see distorted bridge components, designers making inexplicable errors, MnDOT personnel who moved slowly to retrofit an obsolete bridge, contractors who designed and built a new bridge in record time, and congressional representatives who quickly pulled together the votes to pay for a new bridge. On the material side, a varied cast of entities also make appearances: rust and cracks, gusset plates, roadway surfaces, construction materials, and road networks, to name a few. They come together in many different locations: the banks of the Mississippi, the bridge deck high above, the drafting tables of bridge designers, meeting rooms of the MnDOT, and ultimately the halls of the US Congress. Taking a broad, sociotechnical perspective on repair and infrastructure enables us to see how repair plays out across this wide spectrum of people, things, and places.

Infrastructures as Vulnerable Systems

Infrastructure has become an increasingly settled and necessary term for a central phenomenon of modern life: the large, interconnected, standardized technological systems that play an essential role in our everyday activities. However, this use of the term has relatively recent origins that are closely connected to an interlocking set of interests around defense, economic development, and the engineering profession. The concept of infrastructure has historically become relevant when people are concerned about the destruction, sustainability, repair, and maintenance of socially essential systems. With this in mind, the focus on repair and infrastructure in this book is not just an arbitrary convergence of two unrelated topics; in fact, this convergence cuts to the heart of what both infrastructure and repair are all about and why we as a society care about these topics at this historical moment.

The earliest use of the term *infrastructure* in something similar to the modern sense was in the 1950s, mainly in a military context, to refer to the "fixed installations which are necessary for the effective deployment and operations of modern armed forces,"[28] including not only military installations, but highways, bridges, airports, and power plants—all of which were seen as potential targets for strategic bombing.[29] Also in the 1950s and 1960s, the concept of infrastructure gained currency in the field of international economic development, where it was used to describe the "capital base" a country needs to invest in to enable the development of other economic activities. Again, this included things like highways and power plants, but over time, it came to encompass a wider range of systems, including social institutions like education and health care.[30] Building on these earlier uses of the term, infrastructure became more broadly relevant starting in the early 1980s, this time embedded in alarming narratives about the poor condition of US infrastructure, here referring mostly to public works such as roads, bridges, water pipes, and sewers. This usage was tied to economic recession and the argument that investment in infrastructure was critical to restoring economic prosperity. Planning and public works organizations released influential reports with dramatic titles like *America in Ruins* that emphasized this point.[31] Although investment in new infrastructure was encouraged, these reports also documented a dire need for the maintenance of existing infrastructure.

This narrative has proven remarkably durable, in part because of the strategy of assigning grades to US infrastructure, which originated in a 1988 report and has since been taken up by the American Society of Civil Engineers (ASCE).[32] The ASCE has published its version of a national "Infrastructure Report Card" every two to four years since 1998, which has never given US infrastructure a grade higher than a close-to-failing D+.[33] This provides a powerful rhetorical framing that seems to be invoked any time infrastructure becomes problematic. One reason the collapse of the I-35W bridge briefly resonated at the national level was that multiple media accounts attached it to this preexisting, easily understood narrative.[34] This narrative remains relevant as we write this book, having been a major topic of bipartisan interest in the 2016 US presidential election campaign.[35]

Since the 1990s, and especially after the terrorist attacks of September 11, 2001, infrastructure has gained broader relevance, particularly in the United States, in the context of "homeland security" and protection of "critical infrastructure" from natural disasters and attacks.[36] In this role, its meaning has come to encompass all of the sectors mentioned above—from massive engineered systems to complex, changeable sociotechnical arrangements like health care and financial services (see the text box).

As this brief history indicates, the concept of infrastructure encompasses not just a particular set of technological systems, but a central anxiety of modern life: the idea that we increasingly depend on vast, complex, interconnected webs of essential systems that may be unexpectedly vulnerable, placing economic and political stability at risk. In other words, the concept of infrastructure emerged alongside and as part of a larger social discourse on risk, vulnerability, technological decay, and repair and maintenance.

The 2013 US National Infrastructure Protection Plan

A series of executive orders and policy directives by the presidential administration of Barack Obama led the US Department of Homeland Security to develop the 2013 National Infrastructure Protection Plan (NIPP), itself an outgrowth of security policy directives from the George W. Bush administration in the first several years after 9/11. The NIPP identifies sixteen critical infrastructure sectors "considered so vital to the United States that their incapacitation or destruction would have a debilitating effect on security, national economic security, national public health or safety, or any combination thereof."[1] These sixteen sectors provide a useful orientation to the wide range of systems considered to be infrastructure in this context.

The Sixteen Critical Infrastructure Sectors according to the NIPP

Chemical

Commercial Facilities

Communications

Critical Manufacturing

Dams

Defense Industrial Base

Emergency Services

Energy

Financial Services

Food and Agriculture

Government Facilities

Healthcare and Public Health

Information Technology

Nuclear Reactors, Materials, and Waste

Transportation Systems

Water and Wastewater Systems[2]

---

1 US Department of Homeland Security, "Critical Infrastructure Sectors," Department of Homeland Security, March 5, 2013, https://www.dhs.gov/cisa/critical-infrastructure-sectors.

2 US Department of Homeland Security, "NIPP 2013: Partnering for Critical Infrastructure Security and Resilience," 11. https://www.cisa.gov/sites/default/files/publications/national-infrastructure-protection-plan-2013-508.pdf.

Infrastructure in Science and Technology Studies

Alongside the increasing public interest in infrastructure, it has become a core topic in the field of science and technology studies (STS), the primary community we participate in as scholars, and in related fields like anthropology, geography, urban studies, and information studies.[37] Work in all of these fields is diverse, yet united in treating infrastructures as sociotechnical systems, which is also key to our perspective here.

STS takes a broad view of infrastructure that incorporates both its small- and large-scale properties, and the tensions that emerge between these scales. At a larger scale, STS studies have emphasized the structural characteristics of infrastructures, focusing on those that distinguish infrastructures from other kinds of sociotechnical systems. From this top-down perspective, which originated from historical studies of large technological systems, a system becomes an infrastructure when it transcends its regional context to connect with other systems and gains national or global reach.[38] For example, early electric lighting systems served a single neighborhood or city but have evolved into nationally or internationally connected power grids.[39] This kind of global reach is enabled by the development of universal gateways and standards, such as physical plugs and sockets, communication protocols, and classification frameworks.[40] At this scale, structural connections among infrastructure networks, geophysical forces, and governance become more important than local interactions.

STS studies have also examined infrastructures in localized settings where people work and interact. Here, infrastructures are distinguished more by their relationships to other elements of these settings than by their structural features. Specifically, infrastructures show up as integrated resources that support and enable a wide range of local activities and technologies, becoming infrastructures *for* those activities and technologies.[41] This research, which has more sociological roots, represents a kind of bottom-up perspective, focusing on how infrastructures enable and constrain users: for example, something that serves as functional infrastructure for one person or group—such as stairs—may become a barrier to others—for example, people who use wheelchairs.[42] In this view, infrastructures are characterized by the way they are embedded, "sunk into and inside of other structures, social arrangements, and technologies."[43]

There are important connections between the top-down and bottom-up perspectives on infrastructure in STS: the global reach and standardization of large-scale infrastructure systems makes it easier to integrate them into a wide range of localized activities, but the local work of assembling and adapting infrastructure resources to meet local needs shapes demand and spurs efforts toward repair and innovation.

The Growing Interest in Repair and Maintenance

Increasing public concern about breakdown, repair, and maintenance of infrastructure has also coincided with increased interest in these topics in STS and related fields, particularly in the past decade or so. Starting well before this recent surge of interest, our own work played an early role in establishing repair and maintenance as important topics in STS and sociology contexts, beginning with Henke's 2000 article, "The Mechanics of Workplace Order: Toward a Sociology of Repair," a close ethnographic study of repair and maintenance technicians at a university, and his 2008 book on the role of repair in the California agricultural industry.[44] This work, along with related work by STS and urban studies scholar Stephen Graham and geographer Nigel Thrift[45] and sociologist Tim Dant,[46] helped prepare the ground for more recent work that makes a broader case for the importance of repair and maintenance as a topic of academic as well as practical interest, supported by numerous new case studies.[47]

STS and information studies scholar Steven Jackson, writing in 2014, coined the evocative term *broken world thinking* to explain why repair and maintenance must become central to our understanding of the nature of technology.[48] Broken world thinking is ubiquitous in the twenty-first century, which is increasingly "a world of risk and uncertainty, growth and decay, and fragmentation, dissolution, and breakdown."[49] By taking a broken world perspective, according to Jackson, we come to appreciate the tension between "an always-almost-falling-apart world"[50] and the ongoing processes that hold things together and generate productive change. Repair plays a key role in managing this tension: "The fulcrum of these two worlds is repair: the subtle acts of care by which order and meaning in complex sociotechnical systems are maintained and transformed, human value is preserved and extended, and the complicated work of fitting to the varied circumstances of organizations, systems, and lives is accomplished."[51] As we have discussed, infrastructure too is a concept that has its origins in a kind of broken world thinking, and this is reinforced by the always-under-repair character of infrastructures emphasized in the I-35W bridge example.

Jackson goes on to argue that repair is in fact a key aspect of innovation itself, with the development of world-changing technologies like the internet driven not by straightforward processes of design and implementation, but rather by messy processes of implementation, breakdown, problem solving, and repair that over time lead to increasingly stable systems. He also argues that paying attention to breakdown and repair illuminates aspects of technology we might otherwise ignore, in particular the relationships of care and engagement that people have with technology.

STS and history of technology professors Andrew Russell and Lee Vinsel have taken a more activist approach to repair and maintenance issues, organizing a scholarly community and series of conferences under the rubric of "The Maintainers,"[52] and writing about these issues in nonacademic publications like the *New York Times*.[53] In a 2016 piece for *Aeon*, they argue that America has an unhealthy obsession with innovation and "flashy, shiny, trivial things," while we tend to ignore the reality that most technical work and much of the value we derive from technology involves the maintenance of old or unexciting technologies.[54] To counter this tendency, they argue that we should pay more attention to what they call "the maintainers, those individuals whose work keeps ordinary existence going rather than introducing novel things."[55] Their definition of the maintainers extends beyond what we might typically associate with maintenance work, to include those who engage in domestic labor and other forms of care work that are often ignored. They also associate the lack of attention to maintainers with our similar lack of attention to unglamorous technologies like infrastructure, with its continuing need for repair and maintenance. Compared to Jackson, who emphasizes the unexpected continuity between repair and innovation, Russell and Vinsel articulate a more urgent moral agenda aimed at uncovering and promoting the values of care and sustainability they see as going along with repair and maintenance.

Like these authors, we believe that maintenance and repair have a particularly important role in relation to infrastructure and that both of these topics are often understudied and undervalued; that understanding repair and maintenance is absolutely essential to understanding how infrastructures and social institutions can persist, adapt to, and drive change; and that repair provides an essential window into the ethics of care around socio-technical systems. Along with Jackson, we see repair and maintenance as agents of change and innovation as much as conservation of existing arrangements.

Our approach is distinct from these authors primarily in the way we connect repair and maintenance more closely to sociological concepts and theories, in particular by emphasizing the continuity between technological and social order, and the role of conflict and power relations in shaping repair. As a result, we are perhaps a little less interested in making a case for the virtues of the activities of repair and maintenance and more interested in their value as a probe into the complexities and contradictions of our current relationship with technology. We are mindful that repair can serve to prevent needed social change, exclude or remove certain people or groups from positions of power, and stabilize problematic infrastructures and institutions, often at great social cost. However, we do agree that the path from breakdown to repair is a messy, conflicted, and potentially creative process that can also open up opportunities for (sometimes radical) social and technological change. Our approach in this book is also, we hope, particularly concrete and systematic, going beyond establishing why or how repair is important and into developing more specific tools and concepts for understanding and analyzing it.

Alongside repair and maintenance, a number of other concepts also play important roles in broken world thinking. In particular, resilience and sustainability have gained wide currency in many fields as frameworks for recognizing and designing for the inevitability of breakdown and repair.[56] Although we discuss sustainability at length in chapter 5, our overall focus remains on repair and maintenance because these concepts connect more directly to the level of concrete practices—the specific, day-to-day actions people take to pull together social and material resources and get things done. This work is done by people from many walks of life, from well-paid professionals to manual laborers, with different sorts of training and skills, from those learned in college classrooms to those picked up from years of hands-on, on-the-job experience. Indeed, in our definition, repair work is something we all engage in almost daily. By tracing how this work gets done across a broad spectrum of society, we are better able to understand how the stability, resilience, and sustainability of infrastructures are realized in practice, including all the messiness, complexity, contingency, problem solving, and power struggles this can involve. While we, like many others currently writing about repair, ultimately offer sociotechnical repair as a broader framework for understanding aspects of the modern world, we believe that a framework that emphasizes work and practice can provide uniquely grounded insights into the workings of sociotechnical systems.

Repair, Infrastructure, and the Maintenance of Modernity

The increasing interest in repair and maintenance is just one expression of a larger, ongoing societal reckoning with the risks and limitations of science and technology, and their relationship with the human and natural world. The processes of industrialization and modernization that built the world we live in today were largely motivated by the belief that rational, systematic planning and design could provide solutions to nearly every human problem. In this modern ideal, meticulously designed technological systems would ultimately insulate human society from the hazards of the natural world and the tedium of mundane labor.

This vision has proven amazingly productive in some ways but problematic in others. Since at least the 1960s, recognition has been increasing that industrial systems can themselves fail in myriad ways and create entirely new sets of risks their designers did not anticipate.[57] Ulrich Beck, Anthony Giddens, and other social theorists have argued that these aspects of modernity lead to a *risk society*, in which risks are increasingly difficult to anticipate and global in their impact, transcending national borders and social distinctions. This in turn leads to a new *reflexive modernity*, in which belief in technological progress is balanced by attempts to understand and manage the resulting risks to nature and society.[58] One effect of this new modernity is to muddy the sharp distinctions modern planners were trying to build among society, technology, and nature, revealing that they were perhaps never so separable after all.[59] Although much of this literature focuses on the globalization of environmental risks, the increasing globalization of infrastructure systems is equally relevant, tying people around the world to common technological paradigms, with potentially common failure modes.

While these developments are arguably part of the origin of broken world thinking, they tell only part of the story—the part about modern sociotechnical systems inevitably breaking down or creating unanticipated problems. By contrast, most current writers about repair and maintenance take a somewhat more optimistic approach, focusing on the possibility of renewal and sustainability even in a world of constant breakdowns. While we share this sense of optimism, we also see a need for a more skeptical and reflexive view of repair itself to address the role it might play in sustaining some of the very problems pointed out by scholars of the risk society. We return to these concerns in chapter 5.

The Tool Kit

Just as repair workers need a set of tools for their work, we need a conceptual tool kit to help dig up the often-hidden work of repair and maintenance and expose its significance. We focus on three key analytic tools. First, to understand how repair is connected to both things in the material world, and how people think about, talk about, and organize repair of those things, we look closely at the interplay between *materiality and discourse*. Second, to examine how repair can support or threaten cultural and political power structures, and how both infrastructures and those who work on them can become invisible even as they remain essential, we emphasize the relationship between *power and invisibility*. Finally, we examine *scale*—the temporal and spatial scope of infrastructure repair, and how it shapes technological systems in time and space.

Materiality and Discourse

The material work of repair does not take place in isolation. Instead, it is usually accompanied by a great deal of thinking, talking, and writing about what went wrong and how to fix it. Looking at repair through the lens of materiality and discourse provides a tool for analyzing the full range of social and technical relationships that surround repair. In particular, material repair is usually accompanied by discussion and negotiation, however informal, over issues including whether a breakdown occurred at all, what the breakdown was, how to fix it, and what a successful repair looks like. This discourse shapes and is shaped by our engagement with the material system under consideration. For example, a repair technician called in to address a cold office might spend as much time managing the perceptions of the occupants as working on the heating and ventilation hardware (see chapter 2).

In sociology, the most extensive body of work on repair is not about material repair at all, but about what people do to repair shared understanding when confusion arises in conversations. Although seemingly far removed from activities like fixing a bridge or bicycle, this work is a useful starting point for thinking about the discursive elements of all kinds of repair. In the sociological fields of ethnomethodology and symbolic interaction, *repair* refers to the small yet continual efforts people make daily to maintain social order. In conversations, sociologists observe that there are frequent ambiguities that lead to small misunderstandings—you thought "she" referred to Maria, but the speaker was talking about Margaret; you were asking a question, but the other person interpreted it as a statement. People use a variety of interactional strategies to identify these breakdowns and fix them through clarification, redirecting the conversation, and other conversational techniques of repair so communication can proceed uninterrupted.[60]

This perspective suggests social order is an ongoing practical accomplishment, constantly maintained through interaction and negotiation among participants. While we all recognize basic social structures and cultural meanings, the actual everyday enactment of these structures and meanings is highly improvisational. The flow of interactional repair is so ubiquitous that we usually do not pay much attention to it, but we are constantly involved in a dynamic exchange of interactions that create stability as well as possibilities for change. Possibilities for stability and change in infrastructural systems are similarly embedded in a continual flow of maintenance, breakdown, and repair. This flow incorporates both material interventions and repair of meaning in discourse and other forms of social interaction. In fact, these activities are so thoroughly intertwined that it is hard to separate them, so they are best captured through a more general idea of *sociotechnical repair*, which identifies the full scope of material and discursive interventions that make it possible to maintain modern technological systems. Sociotechnical repair includes both routine maintenance and crisis response, fixes to organizations as well as material structures, and a range of scales from local interactions to large-scale institutional processes.

The concept of sociotechnical repair helps make sense of interactions surrounding breakdown and repair in cases like the I-35W bridge collapse. Sociotechnical repair is given shape and specificity through our various accounts of what went wrong and what should be done to fix it. These accounts can invoke a wide range of causes, from material flaws and physical forces to human error and poorly written documents. Repair efforts often go beyond material restoration, to encompass changes in policies, procedures, and communication that serve to restore public confidence in administrators and technical experts. This reflects the thoroughly sociotechnical nature of infrastructures. Their stable material form is inseparable from the array of technical and organizational activities that sustain them. Paying attention to the material and discursive elements of repair helps us map these sociotechnical connections.

While we will not attempt to give a complete account of all possible relationships between discourse and materiality in sociotechnical repair, one thing we have observed is that discourse about repair is often framed in terms of what we call *slippage*, which draws a contrast between a system's current state and some desired state, and poses repair as a way to bridge the gap.[61] One common way of articulating slippage is in terms of change or degradation in a system that renders it unable to fulfill a desired function, as in the case of a bridge that has become too weakened by rust to safely carry traffic. Just as often, however, changes in a system's surrounding circumstances are used to motivate repair—for example, when a bridge comes to be seen as inadequate due to changes in user requirements or engineering standards. This emphasizes the point that breakdowns do not always take the form of something literally falling apart, but can involve much more complex appraisals of functionality in light of many related factors.

Power and Invisibility

Discourses and materiality are inherently tied up in relationships of power; when one or another group of actors controls discourses of repair, for example, they can set the agenda for a sociotechnical system. Those lacking this power can be marginalized, even though they may play an important part behind the scenes for a particular infrastructure. Therefore, understanding the role of power and invisibility in the repair and maintenance of infrastructures is another important element of our analytical tool kit.

Invisibility, or rather tension between visibility and invisibility, is a key theme in infrastructure studies. While infrastructure systems are often intended to be so standardized and reliable that they fade into the background, in other circumstances, they are made very visible, by accident or by design.[62] This tension shows up in several ways in relation to infrastructure and repair.

First, those of us who have access to reliable infrastructures as end users often take them for granted—our use of them becomes so routine across such a wide range of activities and contexts that we may not give them much conscious thought on a day-to-day basis. When infrastructures break down, however, they can become problematic and visible indeed. Think of the household difficulties posed by power outages, or the unexpected and shocking nature of the I-35W bridge collapse, which became an international news story. Other breakdowns emerge more gradually as members of the public, professional communities, or political actors advance agendas for change and repair. And some infrastructures are never fully reliable, existing in a constant state of disrepair. In any of these cases, breakdowns can provide useful insights into the nature of infrastructure, making its more obscure aspects visible in ways that force us to reckon with its materiality and social roles in new ways, a form of what Geoffrey Bowker has called *infrastructural inversion*.[63] Repair and maintenance are aspects of infrastructure that often come to the fore in these situations. Once people believe a breakdown has been resolved, however, these aspects tend to recede into the background once again.

Second, the people who maintain and repair infrastructures are often not very visible to outsiders. In part, this is tied to the general tendency of infrastructure itself to fade into the background. But sociologists also observe that jobs involving "dirty work"—relating to activities, materials, or people society as a whole would prefer to ignore—are often stigmatized and kept out of sight.[64] Repair and maintenance work, particularly when it involves close contact with the material elements of infrastructure, can be "dirty" in two senses. First, it often involves engagement with dirty or dangerous materials, such as oil, dust, or infectious waste. Second, it is associated with breakdowns in material and social orders and may serve as an uncomfortable reminder of the fragility of these orders.[65] Higher-level professionals who manage infrastructure and repair, like engineers, administrators, and planners, may have higher social status, but the nature of their work is often obscure to the general public.

Finally, infrastructures can affect the visibility of certain individuals and communities in social and political life: lack of wheelchair access keeps people with disabilities out of many public places despite recent progress in this area in some regions of the world; freeway bridges and viaducts are often built through neighborhoods with little political power and enable drivers from elsewhere to bypass those communities (see chapter 3); and some rural areas and urban neighborhoods are marginalized by limited access to internet services.[66]

All of these issues with visibility and invisibility reflect the larger connection between infrastructure systems and systems of cultural and political power. When people are marginalized in relation to infrastructures, they also tend to be subordinated within power structures; conversely, infrastructure projects often enhance the power of states and social elites. Infrastructures enable large-scale access to and movement of natural resources, support economic activity, and serve as symbols of administrative competence and control over state territory.[67] In this role, they are subtle reminders of the ubiquity of state and institutional control, as we discuss in chapter 4. And some elements of infrastructure are anything but invisible, instead taking on the status of civic or cultural icons. Think of the aesthetic qualities and visual prominence of major bridges, train stations, and airports, or the way certain cars or cell phones function as status symbols.[68] These dynamics of visibility and invisibility of both material structures and communities play a key role in the Coronado Bridge case study presented in chapter 3.

When infrastructures break down, they can threaten the stability of systems of power, as well as provide opportunities for marginalized groups to gain power. For this reason, infrastructure repair is often as much about restoring or extending systems of power as it is about restoring material order—hence, the political urgency of replacing the I-35W bridge. At the same time, it can be a location for power struggles and renegotiation of social and material orders.

One useful way of understanding the relationship between repair and power is by distinguishing two general approaches to repair: *repair as maintenance* and *repair as transformation*. The goal of repair as maintenance is conservation of the status quo: protecting or restoring a preexisting set of practices, relationships, and power structures. Groups and individuals who benefit from the current sociotechnical order are likely to support this kind of repair, particularly if it helps them hold their position in systems of power. Our use of the term *maintenance* here encompasses a lot more than routine maintenance and upkeep; we also include more focused repair efforts in response to unexpected breakdowns both large and small as long as the goal is a return to the status quo. The replacement of the I-35W bridge, along with associated changes in bridge design and inspection procedures, is a good example of repair as maintenance. Despite the extraordinary effort involved, its results were not exactly transformative: it restored the local transportation network to its former configuration, while preventing a potential breakdown in public confidence in powerful groups, including city and state leaders, government agencies, and the engineering profession. No wonder it was agreed to, funded, and completed so quickly.

Sometimes, however, repair is presented as an opportunity for more radical change in existing structures and practices. Transformative repair may serve the interests of the powerful when some fundamental breakdown has occurred that permanently threatens their place in the current order. Such was the case in the US nuclear weapons complex at the end of the Cold War. With no prospect of maintaining the infrastructures that enabled the building and testing of new nuclear weapons, weapons scientists were able to construct a stewardship role for themselves that leaned heavily on a new modeling and simulation infrastructure. This fundamentally changed their role but maintained their professional status and credibility (discussed in more detail in chapter 4).[69] But transformative repair may also be championed by those who are excluded from existing power structures or otherwise see repair as an opportunity to advance their interests. This was the case with the community activists who became involved in the Coronado Bridge retrofit project described in chapter 3.

To summarize, understanding how an infrastructure system distributes risk, benefits, and social status unevenly across different groups and individuals can help us understand how it intersects with systems of cultural and political power. This, in turn, can make it easier to understand what is at stake in its repair: sometimes the stakes are large, sometimes small; sometimes there is broad agreement about a course of action, sometimes conflict driven by who gains and who loses in repair. Proposed repair strategies can reflect not only immediate material and technical needs, but also the interests of stakeholders in maintaining or changing the existing social and material order.

Scale

Close examination of the scale of repair in both space and time is a tool that enables us to analyze how repair works on infrastructures at a range of social and material levels, from our daily local practices, to the national politics of infrastructure, to management of global systems of exchange. While many studies of repair focus on local practices, the temporal and spatial extent of repair activities can vary tremendously across these levels. A focus on scale ensures that we do not miss any important aspects of materiality and discourse, or power and invisibility, due to preconceptions about the scope of repair. In keeping with the importance of scale, this book is organized in chapters that address successively larger scales of repair, from local negotiations to systemic, national, and global interventions. This should not be taken to imply that these scales are necessarily distinct from each other in actuality, and we emphasize interactions and continuities across scales throughout.

Questions of scale are important in infrastructural repair in part because infrastructure systems occupy intermediate scales between the human body and the geophysical world in terms of size, temporal horizons, and ability to exert force. This connection between scales can serve as a powerful amplifier of human action and ability to control the natural environment.[70] By the same token, dependence on infrastructure creates new sources of vulnerability, because when natural forces do overwhelm infrastructure systems, the effects can reverberate back down to the scale of fragile human lives. As a result, most natural disasters in the modern world are simultaneously infrastructural disasters.[71] In short, both infrastructures and nature now function as increasingly integrated parts of a larger material environment that both provides us with resources for daily living and presents dangers we must protect ourselves against.

One way researchers in STS and related fields have managed issues of scale in the study of infrastructure is by considering it from both top-down and bottom-up perspectives, the first emphasizing the large-scale structural characteristics of infrastructure systems, the second focusing on their role as a resource supporting local technological and social arrangements. These two perspectives also suggest useful ways of analyzing scales of repair.

From a bottom-up perspective, repair emerges from a complex tangle of relationships among material objects and systems, social relationships, and embodied skills. In the bottom-up view, repair is mainly about negotiating and renegotiating local forms of technological and social order. Examples might include a mechanic fixing a car, a technician adjusting ventilation airflow in an office, or a nurse repositioning a patient to minimize pain.

One key aspect of repair at the local level is the relationship between the repair worker's body and the material world. When we engage in diagnostic or repair work, this connection is readily apparent: we use our hands and body to physically manipulate the thing being repaired, usually with tools as an intermediary; position our eyes to get different views on the repair; feel for vibrations and temperature; listen for unusual sounds; and so on. Another key aspect of local repair is the way it fixes human as well as material relationships. A successful repair is not only a material accomplishment; repair also creates a shared narrative of what went wrong and how it was fixed, persuading participants that local sociotechnical order has been restored (see chapter 2).[72]

From a larger, systemic point of view, the work of repair can extend across multiple locations, institutions may take a larger role than individuals, and repair may be more extended in time. Examples include the rebuilding of levees and other infrastructure after Hurricane Katrina devastated New Orleans,[73] or the reconfiguring of nuclear weapons infrastructure following the Cold War.[74] From this perspective, the emphasis may shift from hands-on repair workers to professional managers, planners, and engineers; the organizations that employ them; and governmental and professional decision-making bodies. These entities engage with the social and material aspects of infrastructure on a larger and more abstract scale, which might involve managing the movement of large quantities of material, purchasing and deploying large construction equipment, or interacting with political actors and the general public. Much like infrastructure itself, repair at this level can be regulated through standards and common frameworks.[75] Discourses and narratives of repair are still important at larger scales, but may be constructed in more public ways, including through media reporting, public meetings, social media, and the political process. It is at this level that questions of institutional and state power become most noticeable.

We talk about scale here in terms of perspectives because, in reality, what we see when we look at different scales of repair are just different aspects of a single phenomenon in which sociotechnical systems are reconfigured and adapted to changing circumstances. Infrastructure systems, in particular, tend to span multiple scales, so we need to look at them from different perspectives to put together a complete picture of repair. This perspectival approach to repair also helps us avoid the pitfall of equating scale with hierarchy or suggesting that some levels have more importance than others.[76]

But while it is important to understand the continuities between different scales of repair, it is also important to recognize discontinuities where they do occur. One source of discontinuity is the fact that infrastructure systems are often deliberately structured in modular or layered configurations as a strategy for managing their complexity.[77] For example, roads and bridges are pieces of infrastructure that are embedded in very specific local settings and activities, but at a larger scale, they are components in a road transportation infrastructure that includes interchanges, traffic lights, cars, trucks, and drivers. This larger infrastructure can experience breakdowns even if the bridges and roads remain intact—for example, due to unexpected traffic volumes or major car accidents or failures in lower-level components like the I-35W bridge that can also damage the larger system. These different layers of infrastructure can be distinctive both socially and technically; they may be managed by different groups of professionals, may employ distinct technologies, and may evolve somewhat independently despite their interconnections. For example, the replacement of the old, fracture-critical I-35W bridge with a new, high-tech bridge represented a leap forward in bridge technology, but by design, it had a limited impact on the overall transportation system in the Twin Cities area. Because of these distinctions, repair practices can also evolve somewhat independently within different layers. We may find, for example, that bridge engineers and bridge construction and maintenance workers are distinct communities with different perspectives that have to be reconciled when new kinds of repair become necessary.[78]

The temporal scale of repair is closely related to its spatial scale.[79] Many smaller, local repairs take place within short, well-defined time limits. Rebuilding after a disaster like the I-35W bridge collapse may take longer, but can be planned as a one-time event and without much concern for the overall life cycle of infrastructure. More systemic repairs can encompass a wider range of timescales: restoration of a levee system after a hurricane may take a few years, while restoration of housing and street life in flooded neighborhoods might take decades and connect to long-term economic, demographic, and political trends. Concepts like routine, preventive, or life cycle maintenance envision repair as an ongoing process throughout the life span of a technological system. Sometimes repair work never really restores a system to a desired state, and continues indefinitely in a state of incompletion (see the text box in chapter 4).[80]

The temporal relationship between breakdown and repair can also be important. While it might at first seem that all repair must take place after a breakdown is identified, this is not always true. Preventive maintenance, for example, might mean monitoring a system for signs of approaching failures and fixing the system before they can occur. Breakdowns can increasingly be anticipated through experiments, tests, and computer simulations, enabling planning for repair of systems that have not even been built yet.[81]

These ways of thinking about spatial and temporal scale are important tools for analyzing repair because they allow us to more easily trace the full scope of repair and its connections to what comes before and after. By carefully tracing connections in space and time, apparently simple cases of repair can reveal themselves to be much more complex, interesting, and challenging than they seem at first glance. Indeed, one of the most important aspects of repair may be the way it serves to continually connect and reconnect sociotechnical systems across space and time, which is part of what makes the characteristic scope and persistence of infrastructure systems possible.

Structure of the Book

The remainder of this book is structured around the scale of infrastructural repair. We begin in chapter 2 with the local context of repair, focusing on how people negotiate the social and material aspects of repair in specific places. We introduce the concept of the networked body to describe how the capacities and senses of the human body serve as a crucial link between the material and the social in repair. We explore this concept through a number of case studies, including Henke's studies of university maintenance workers. We conclude by examining the connections between repair and maintenance work and care work, and examining the emergence of local repair collectives in various global settings.

In chapter 3, we move up in scale to examine interactions between local negotiations and systemic considerations in repair of infrastructure systems. This chapter focuses on several episodes from Sims's work on California's efforts to retrofit its bridges to protect them from earthquakes, with a focus on how this systemic repair effort collided with local concerns surrounding the San Diego–Coronado Bridge. We explore how the construction of this bridge devastated the Barrio Logan community in San Diego and how the community fought back to repair this intrusion by taking control over the area under the bridge to build a community park, incorporating a series of monumental murals painted on the bridge columns. We then examine how engineers in California came to recognize that bridges in the state were not adequately designed to resist earthquakes and how a community of state engineers and academic researchers came together to develop a program for seismic retrofit of bridges. Finally, we explore how these professional communities negotiated with the Barrio Logan community to find a retrofit solution for the San Diego–Coronado Bridge that would not damage the murals. This chapter emphasizes the potential complexity of power relations around infrastructure and repair—in particular, how the connections infrastructure makes between marginal and highly valued urban spaces can open up possibilities for disenfranchised communities to challenge established power structures through processes of repair.

Chapter 4 focuses on how repair plays out through large-scale relationships between experts, infrastructure, and the modern state. It shows how, beginning in the seventeenth century, infrastructure increasingly became an engine for developing state power and reinforcing national identities. In particular, infrastructure projects became important tools for accumulating capital under elite control, while also serving as highly visible symbols of the capability of the state and its ability to serve its citizens. In these roles, infrastructure became a crucial instrument of governmentality, placing expertise and technical knowledge at the heart of modern state formation. The chapter then draws on our previous work to explore how these issues came into play in the efforts to preserve US nuclear weapons knowledge and infrastructure after the end of nuclear testing and the Cold War. This case study shows how experts and elites can mobilize to repair systems that give them power in ways that preserve that power, sometimes even at the cost of massive transformations in how these systems operate.

Finally, in chapter 5, we turn to a global scale, examining the place of infrastructure and repair in the Anthropocene, a new geological era that has been proposed to describe the increasingly global impact of human activities on the natural environment. Infrastructural systems are arguably the key drivers of the changes that mark the beginning of the Anthropocene, whether we trace those changes to the beginnings of systematic agriculture and its impact on natural ecosystems, the development of massive carbon emissions generated by transportation and energy production, or the radioactive traces generated by nuclear weapons testing and production systems. This raises concerns about the ethics of repairing systems that may ultimately have adverse and unsustainable effects on the natural world and human health and prosperity and leads to a critical question: Can we repair infrastructural repair itself to ensure that we have infrastructural systems that are not just stable but resilient and sustainable in the long term? To explore this question, we examine how infrastructures became increasingly globalized in the course of the twentieth century, how global governance structures have emerged to regulate these infrastructures, and how understanding these impacts requires massive information and data-gathering infrastructures. We close with a discussion of different visions for the future of infrastructure and repair and some guidelines for practicing and analyzing repair reflexively, acknowledging complexities and contradictions in the relationships between humans, infrastructures, and the natural world in the Anthropocene.

2 Cold Offices and Hot Airplanes: Local Negotiations over Repair

The Cold Office Problem

An office administrator makes a call to the maintenance center at the university where she works. Her complaint? The office is too cold, and could someone check out the problem for her? When the mechanic arrives to investigate the office's ventilation system, he finds that there is "no problem"—the office temperature and the rate of airflow through the office vents are within the range of normal expected for that building and day. While the mechanic used instruments and flashing numbers to convince the office worker that her workplace was not in need of repair, she still felt cold and was frustrated by the lack of a straightforward fix for her complaint.

This brief encounter, taken from Henke's research with a group of university-based physical plant mechanics, describes what we term the "cold office problem" and is likely a common situation for workers in offices all around the world (especially where temperature control—or even opening a window—is not at the discretion of the workers inhabiting the office space). The cold office problem raises a lot of interesting questions about repair. How do we really know when something is broken? Who gets to decide if something needs to be fixed, the best approach for fixing it, and the criteria for deciding when a fix is complete? What happens when people disagree about repair? Does the status or identity of the persons involved shift the process of mending things? Not all cases of repair involve these kinds of questions, but they are the ones that interest us here, and they also tend to be pretty common. Anyone who has taken in their car for service, who has been to the doctor to have a bodily complaint checked out—or, indeed, anyone who has complained that a room is too cold or too hot—knows that repair can be messy. We usually hope for an easy diagnosis and fix, but it does not always work that way.

In this chapter, we explore the ways that repair is negotiated: subject to conversations, power struggles, and trial-and-error experimentation that shape the process and outcomes of repair work. Our focus here is on the local scale of repair, where hidden infrastructures like office heating and cooling systems intersect with our everyday routines. If repair is negotiated between people and the material stuff that surrounds them, then beginning at this local level of analysis provides a foundation for the coming chapters that examine infrastructure and repair at larger scales.

Exploring repair at the local scale helps us see how infrastructures are negotiated through two elements of the conceptual tool kit we introduced in chapter 1: *materiality and discourse* and *power and invisibility*. The cold office problem illustrates each of these concepts and their connections. Air, duct work, tools, and workers' bodies provide a material basis for the action in this work setting, but at the same time, it is the exchange of cultural meanings and frames through language that allows the two office workers to negotiate whether a room's temperature should be treated as normal or somehow in need of repair.

These interactions both reinforce and are supported by power relations that shape the interactions among various people and things. The cold office problem features a mechanic who has the power to control aspects of the ventilation system for which the office worker is not allowed access. Although the office worker may have access to other sources of power and influence, her lack of control over the air-conditioning system limits her ability to influence the outcome of this particular repair process. It is also important to emphasize that the mechanics at this university are overwhelmingly men, while most of the office administrators are women; the gendering of repair work, and its consequences for repair outcomes, is a topic we return to later in this chapter.

Negotiated repair, then, never takes place in a vacuum. The interactions surrounding it are heavily influenced by preexisting social hierarchies and power relations. This means that some people ultimately have more visibility and influence over repair processes and outcomes than others, though the question of who wields power and who becomes marginalized or invisible is not always easily resolved. An office administrator may have access to organizational decision makers who have influence over hiring and firing of maintenance workers, for example, and in general people who are excluded from certain kinds of power may find alternative ways of exercising control.

Defining and Controlling Thermal Comfort

How is temperature controlled in an office environment? And who gets to set it?

Building engineers have been working on answers to these questions over the past half-century, and one name stands out in the academic literature and engineering practices on thermal comfort in buildings. The Fanger thermal comfort model was developed in the 1960s by Danish engineer Povl Ole Fanger, who performed a number of experiments on human subjects in climate chambers to assess perceived comfort on a seven-point scale (figure 2.1). Correlation of measures such as temperature and airflow with the subjective reports of comfort from test subjects allowed Fanger and colleagues to determine a zone of comfort where the "predicted mean vote" of participants who would likely feel too warm or too cool could be minimized.[1] Fanger's 1973 article, "Assessment of Man's Thermal Comfort in Practice," outlines a number of criteria that can have an impact on feelings of comfort, such as the type of clothing worn (or "clo-units," ranging from naked [0 clo-units], to bikinis [.05 clo-units], to business suits [1 clo-unit]), age, type of environment in which a person was raised, and gender. Fanger included equal numbers of men and women in his experiments and found they "seem to prefer almost the same thermal environments."[2]

Essentially, the Fanger model tries to balance out the thermal preferences of a group of people, and more recent work debates whether this compromise is an appropriate way to set temperature. Neutral values are not always consistent, even from the same person on a day-to-day basis, and may vary in many ways, including seasonally, cross-culturally, and between laboratory and field studies.[3] A letter published in *Nature Climate Change* in 2015 challenged one key aspect of the Fanger model, the metabolic rates of men and women, claiming that the model may overestimate female metabolisms and thus provide temperature settings too low for women. A report on this study in the *New York Times* featured several stories of women who wore heavy sweaters and blankets when working in their office spaces.[4]

This question of how temperature is set is related to but also different from the question of control: Who has control over the systems that provide cooling and heating to a particular building and its living and working spaces? On this point, Fanger and his more recent critics agree: models of thermal comfort work best when their subjects have some discretion over temperature.[5]

Defining and Controlling Thermal Comfort (continued)

In the office building where Henke works, employees do not have direct control over the temperature in their offices; each office has a sensor on the wall that reports air temperature back to a central control system. Those who feel that their office is too cold often place a chilled, damp paper towel over the climate sensor on their wall, to trick the system into pumping out more heat (figure 2.2). It works, giving them partial discretion over a system that is controlled elsewhere on the campus. The paper towel trick illustrates how negotiations between people and things, and technologies and institutions, lead to complexity and fluidity in repair. Order and change, normal and broken, depend in part on one's perspective.

---

1 P. O. Fanger, *Thermal Comfort: Analysis and Applications in Environmental Engineering* (Copenhagen: Danish Technical Press, 1970).

2 P. O. Fanger, "Assessment of Man's Thermal Comfort in Practice," *British Journal of Industrial Medicine* 30 (1973): 320; Fanger, *Thermal Comfort*, 86–87.

3 J. van Hoof, "Forty Years of Fanger's Model of Thermal Comfort: Comfort for All?" *Indoor Air* 18, no. 3 (2008): 182–201; Liu Yang, Haiyan Yan, and Joseph C. Lam, "Thermal Comfort and Building Energy Consumption Implications—a Review," *Applied Energy* 115 (2014): 164–173.

4 Boris Kingma and Wouter van Marken Lichtenbelt, "Energy Consumption in Buildings and Female Thermal Demand," *Nature Climate Change* 5 (2015): 1054–1056; Pam Belluck, "Chilly at Work? Office Formula Was Devised for Men," *New York Times*, August 3, 2015, https://www.nytimes.com/2015/08/04/science/chilly-at-work -a-decades-old-formula-may-be-to-blame.html; S. Karjalainen, "Thermal Comfort and Gender: A Literature Review," *Indoor Air* 22, no. 2 (2012): 96–109; Caroline Criado Perez, *Invisible Women: Data Bias in a World Designed for Men* (New York: Abrams Press, 2019).

5 Fanger, "Assessment of Man's Thermal Comfort in Practice," 313; van Hoof, "Forty Years of Fanger's Model of Thermal Comfort," 182.

At the heart of local repair are human bodies, the material world they inhabit and interact with, and the negotiations among the people, tools, and infrastructures that define the boundaries of repair encounters. The cold office problem, for example, concerns two workers in a specific, material setting—the office and its surrounding infrastructure—and how they negotiate definitions of breakdown and repair based on the sensations and experiences internalized in their bodies and the politics of their own identities and roles.

Figure 2.1
An image from one of Fanger's climate chamber studies, captioned: "Subjects wearing a standard uniform (0.6 clo) during thermal studies in the environmental test chamber at the Technical University of Denmark. During the studies, physical and physiological measurements are correlated with the subjective evaluations of the subjects." Reproduced from P. O. Fanger, "Assessment of Man's Thermal Comfort in Practice," *British Journal of Industrial Medicine* 30 (1973): 319. With permission from BMJ Publishing Group Ltd.

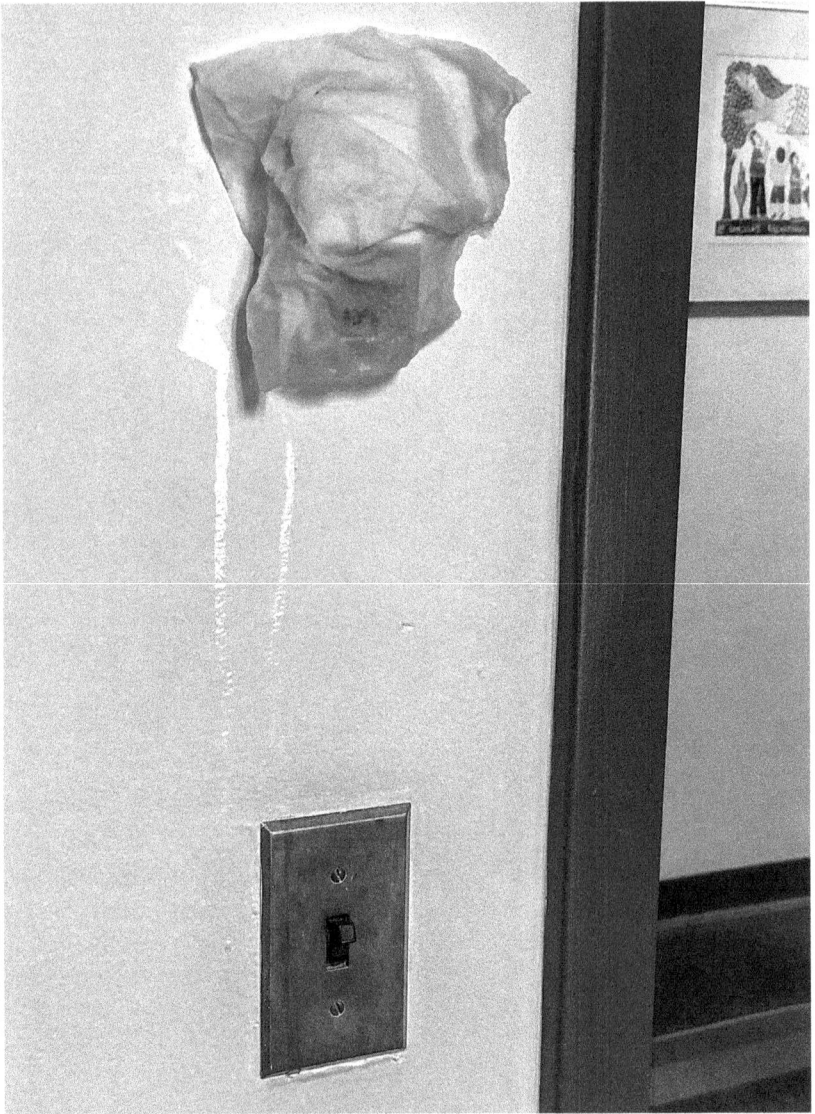

Figure 2.2
The paper towel trick in use. Photograph by Christopher R. Henke, 2018.

In this chapter, we examine several aspects of local repair negotiations. First, we explore the importance of the body and its networked connections to the material world of infrastructure. Next, we look at the connection between materiality and discourse by examining talk about repair and how these conversations serve to negotiate repair in everyday work settings. Talking about repair defines the boundaries of repair communities, as well as the disagreements and conflicting interests that make up the local politics of repair. Finally, we discuss the role of difference in repair, especially the gendering of care and emotional labor, which influences our views about who does what kind of repair and how they ought to do it. Following these differences leads us into further discussion of the role of power and invisibility in repair.

Infrastructures as Local Structures: Bodies and Materiality

A cold office is a local problem. A worker who rubs her hands and arms to warm up feels in her very bones that something is not right with the temperature in her office. This interface of bodies and the material environment they inhabit is a good place to start understanding how repair plays out on a local scale. In later chapters, we examine how infrastructures increasingly stretch around the world and connect us in myriad ways to global systems of communication, travel, and ecological change. Despite this global reach, we typically interact with infrastructures in a very local way. When our car hits a pothole in the road or tainted hamburger meat makes us sick, we feel these events most directly as a jolt to the body, even though the conditions that led to these events are connected to broader technological and political systems.

There is an immediacy and sensual connection in our local relationships with infrastructures that has important consequences for how we understand and act through them. That bodily connection in turn provides an important context for repair, shaping how we identify and negotiate the need for it and our preferred solutions. There are aspects of infrastructure and repair we come to know through our local connections to them that we cannot access or understand in the same way from a distance. So the local work of repair is defined in part by proximity: repair of specific systems, in particular places, accomplished by people working alone or closely together, interacting and engaging with a broken machine or facility. But the materiality of this engagement is also crucial. The stuff we hold to hand and bump against provides us with both abilities and constraints that shape and delimit our actions.

Sociologist Tim Dant describes this relationship as "material interaction," which is a subtle way to describe a profound condition.[1] We exist in a material context, and if that seems obvious, that is in part because social scientists have sometimes eschewed the material in favor of placing more weight on dematerialized "social" explanations.[2] Human actors design infrastructures in specific ways and for specific purposes. Using examples from his work on automobiles and their repair, Dant describes a threaded nut and bolt as "[artifacts that] embody the intentional actions of prior human beings," thereby building meanings and uses right into the form and structure of these objects.[3] In this way, people invest materiality with specific intentions and interests, and those designs inform the actions of other users. As a result, when a mechanic needs to repair a broken nut or bolt, a range of options may be possible, but a replacement of the same size and type of part is likely the most straightforward repair. The broken bolt might be repurposed for another use—even as a piece of art—but the shape and structure of the artifact suggest a specific use. Some scholars of materiality go a step further, emphasizing the ability of so-called nonhuman entities such as other organisms or things to act in ways that might either parallel our own interests and actions or resist them.[4] This nonhuman agency can also be seen as a more passive kind of push or pull, as in Bruno Latour's example of a speed bump, a feature of a roadway that forces drivers to slow down by design, thereby acting on drivers and their intention to speed along without delay.[5]

Latour claims that materiality represents the "missing masses" that are required to give a complete picture of the connections among nature, technology, and human actions and relationships. Materiality is "missing" because it is so ubiquitous that we sometimes do not fully acknowledge its significance.[6] This materialist perspective leaves some questions about the relationship between human and material agency. As the cold office problem demonstrates, people can, and often do, have very different perspectives about the state of material infrastructures.

How can materiality be both a hard set of constraints and something we interpret and act on in flexible ways? We suggest that repair is one of the missing practices that enable us to resolve this tension by negotiating how the social and the material come together in specific places, with human bodies as the crucial link between the two. Local interactions with infrastructures involve repair on a continuous basis, in ways big and small, and the material configuration of infrastructures suggests possible ways of approaching repair when breakdowns occur. For example, if we are having car trouble, the hood of the car provides a natural point of access to the engine, and a modern engine itself incorporates visual cues on how to interact with it that even people without mechanical skills can understand: colored lids to unscrew, dipsticks to pull out, and markings showing acceptable fluid levels. These structures reflect human designs and interests, however, and we may interpret them in disparate ways. An engine provides more and different points of access to an experienced mechanic, who consequently may see very different possibilities for repair than a naive consumer does. Even with their differential expertise, mechanics and car owners may have legitimate disagreements on what kind of repair is necessary—or what a reasonable cost for repair is. This is where seeing repair as a practice of negotiation, between human actors as well as between people and things, helps visualize the ongoing and messy process of working within local contexts of infrastructure and culture.

Consider another example from Henke's work with the building mechanics: a maintenance worker named Al is dispatched to the campus teaching hospital, where the staff are concerned about a burning smell in their surgical ward. The following field notes describe how Al, using his bodily senses, works through the materiality of the hospital's infrastructure as he seeks the source of the trouble:

Al and I respond to a complaint from the medical clinic that there's "a burning smell" near a surgery area. Upon arriving at the scene of the complaint, we can immediately smell the odor as well. The scent is strongest in a hallway just between the reception area and patient recovery room.

Al remarks that the odor smells like a burnt-out ballast—an electrical part of the fluorescent light. Are any lights not working? No, all the lights in this hall area seem to be lit.

Al sets up [a] ladder in the hallway and removes a panel of the white foam ceiling, poking his head above and sniffing. "The scent's not so strong above the ceiling," Al says. "It might be coming from this vent duct."

There is a duct for a ventilation shaft that comes down and ends in a screen in the ceiling, very close to where Al has just sampled the air with his nose. Is the odor coming from the shaft?

The shaft [leads to] an air handling unit on the roof of the building, a device which passes air over hot or cool water—depending on whether the building needs to be heated or cooled. Al and I go to the roof. The air handler is the size of a comfortable office. He opens a door on the unit and again inhales deeply. There is no burning smell.

We return to the hallway area where Al looks above the ceiling again. A nurse observes that half of the EXIT sign is unlit—maybe that's the problem. The other nurses and orderlies standing idly by chuckle at her suggestion. Al moves his ladder to the sign and removes a plastic cover. Sure enough, it was a ballast in the EXIT sign that had literally burned out. It was a blackened coil of wire that reeked of the same smell we encountered when first arriving.[7]

Observing a skilled repair worker like Al in his element, it is hard to miss the role of his body: he uses it immediately to diagnose the source of the burning smell when he arrives on the scene, though it took a sharp visual observation from the nurse to help identify the particular light fixture causing the smell. Al's use of his nose as a tool of repair is an important form of materiality, a key means of connecting his body with the infrastructures that surround and structure his workplace. The skills and knowledge of an experienced mechanic may seem mysterious or even magical to those who observe or depend on repair, as their very embodiment can make it difficult for outsiders to observe how they operate. Embodied repair is paradoxically both visible and invisible: Al's embodied skill—and especially his nose!— were prominent in this episode, but his intuition about the source of the smell was buried in his nose and brain. Thus, the invisibility of infrastructures stems in part from bodily interactions with materiality, as we may not always observe or appreciate the hidden work and experience involved in infrastructural repair.

Another example of the embodied nature of repair work comes from Douglas Harper's book *Working Knowledge*, an ethnographic study that focuses on a mechanic named Willie who owns a small repair shop in rural New York where he fixes cars and farm equipment.[8] Willie is an important person in his community due to his expertise in *bricolage*, or the art of using materials at hand in creative and skilled ways. For Willie's customers, paying cash for a new car or tractor part is often difficult, and so Willie is a master of fabricating and improvising solutions that are less costly and often more effective than replacing old parts. Harper emphasizes the importance of bodily knowledge for this work, as Willie must use craft knowledge of metalwork, welding, and reassembly to complete repairs.

In one part of the book, Harper focuses on Willie's expertise with automobile transmissions, noting that transmissions, purely mechanical devices, should be relatively straightforward to repair because they fit together in a manner governed by the logic of gears and illustrated through technical manuals. And yet: "Even in this most 'objective' of procedures it is the subtle play of force and pressure, the simultaneous movement of parts, and an evaluation of wear through the sensations of the fingertips that guide and control the process of work."[9] At one point when discussing a particularly complex transmission, Willie describes the use of bodily sensations as a connection between hand and mind, where "it's just like your fingers got eyes."[10] Similarly, one of the university mechanics Henke observed described an instrument for testing airflow as allowing him to "look" inside a duct.[11]

While embodied knowledge is a key aspect of Willie's work, focusing just on his bodily skills tells only part of the story. Willie's use of his body in tandem with material objects and infrastructures means that while embodiment is clearly a key element of his repair work, his knowledge is also dependent on the specific context of the repair. Bodies may have what it takes to do repair work, but the work itself ties bodies together with a wider set of materials and discourses. Henke terms this body-in-context the *networked body*, "a body [that] has become situated within a larger setting of activity, rather than simply internalizing a previously external skill."[12] When Willie repairs a transmission, his ability to use his fingers and brain to feel his way through the problem is inherently tied up with the structure of the machine itself, making it hard to disentangle his knowledge from the materiality of the transmission and all its interlocking parts.

This view of embodiment helps us better understand repair in a few ways. First, the network model emphasizes the interconnectedness of diverse groups of people, technologies, and cultural meanings through infrastructures; our bodies get connected to each of these elements through experience and negotiation. Knowledge is not just in the body but distributed through the stuff of particular machines and systems. This leads to a second point about the time-dependent nature of bodily engagement with infrastructural networks. It takes time to understand infrastructures at the level of working knowledge and repair. Networks exist in particular times and places, meaning that while some repair skills might transfer to other contexts, some may not. Third, the networked body reorients the relationship of the body, materiality, and repair, emphasizing that knowledge of infrastructures is not contained just in the bodies of repair workers; it is also embedded within a surrounding network of people and things. In the cold office problem, the office worker *knew in her body* what she perceived to be an abnormal temperature for her office space. While she may not have possessed working knowledge of her office's heating and ventilation system, her body had been networked into them enough that she felt something was wrong and needed to be fixed. This in turn made her an important element in the mechanic's network. Similarly, in the example of the burned light ballast in the hospital, a nurse was the first to notice and point to the fixture of her daily work environment that turned out to be the source of the trouble.

Summing up these points, the networked body helps us understand how we identify situations where repair is needed or necessary. The bodily sensation that something is locally wrong with infrastructures is frequently the genesis of repair, whether this sensation is experienced by a repair worker with intimate knowledge of a system or an everyday user who may not have deep understanding of the underlying problem. Repair, then, is a fundamentally experiential practice and sense of being, a form of awareness that comes from life lived in specific, local settings and in interaction with material systems and bodily sensations. Even when infrastructures are "inverted" and brought to the surface of our awareness through breakdown and trouble, we may not see how repair is applied to them, because the relationships of body, sensation, and structure are subtle and often themselves embedded within the networked body. Repair work is often described as invisible, but it may be more accurate to say that it is inscrutable in action— hard to discern because of the embodiment of infrastructural knowledge and repair skills.

Negotiating Discourses about Infrastructural Repair

This mix of bodies and objects, each providing a different point of view on the status of local systems, makes the study of repair a fruitful approach to understanding the negotiation of infrastructural order and change. In this section, we show this negotiation in more detail, especially the way that *discourse* about repair provides the social and material background for everyday conversations about infrastructures. Discourse is commonly associated with talk and conversation, but social scientists use the term more broadly to include "an interrelated set of 'story-lines' which interprets the world around us and which becomes deeply embedded in societal institutions, agendas, and knowledge claims."[13] Discourse is dynamic in this view, with multiple discourses competing for relevance or becoming attached to particular situations. Discourses are a kind of cultural infrastructure in themselves, supporting how we think and talk about all kinds of things, including materiality. When we discuss and debate the status of infrastructures and the need for repair, discursive stories help us work out our relationships and problems with the material world. Control over stories can also provide a significant source of power for those who argue for one or another approach to repairing infrastructures, as we will see in chapter 3. Tracing discourses around repair is especially fruitful when disagreements occur. What happens when people tell different stories, with competing suggestions for repair or views on whether repair is required at all? Because infrastructure breakdowns create situations of disruption and unease, negotiations around repair may reveal competing narratives of what is wrong, who is at fault, and what should be done. This is where repair becomes a truly negotiated order, with participants giving competing accounts of trouble and what to do about it.

Once again, the cold office problem provides a useful set of examples. In the excerpt of an interview that follows, Henke asks Al, a specialist in heating, ventilation, and air-conditioning (HVAC) systems, to demonstrate some of the tools he commonly uses to do his work. Al describes his use of a tool called a hood flow meter to measure the airflow rate from a duct:

Al: [The flow meter] lets the customer know and see, that you're doing something and that there is actually [laughing] air coming out . . . and you can actually show 'em the reading. It seems to settle 'em down a lot of times. You know that's why in some of those labs you'll see, just a little strip of paper taped off of [the vent]— for years they've been having problems and [saying], "I'm not getting any air!" [mocking customer]. You'll hang that up there and take a reading and . . . let them see that there is air moving, and then you won't get the same repeat calls again.

CRH: 'Cause they can see for themselves.

Al: They can see yeah, and then they'll believe you. Where if you just go in and [hold] your hand [next to the vent] and go, "Yeah you're getting air out of here lady," they don't buy that, they wanna see something with some numbers flashing on it.[14]

Al describes his strategic use of the flow meter to nudge office workers back to a sense of normalcy, to prove via a technical display that air is flowing as it ought to be. Sociologists have noted that when we share a definition of reality in a given situation, we have a discourse for describing and thinking through the activities that make that reality happen.[15] In this case, Al uses the readings from the flow meter to suggest that the customer is mistaken about the reality of the situation in her office and that she should trust the meter as an objective indicator of the truth of his definition of the situation.

Infrastructural Reality

Reality is reality, right?

Many infrastructures have a solidity based in their material form. Pipes and wires, concrete and steel—this is the stuff of a material reality that is hard to deny. And yet as we saw in the case of the cold office problem, two reasonable adults could quite easily disagree about the status of a particular infrastructure and the properties of a specific material environment, including whether the air was comfortable or not. Similarly, in the I-35W bridge example discussed in chapter 1, engineers did not "see" warped gusset plates as a sign of impending structural failure until after the bridge's collapse. Thinking about infrastructures as institutions can help us understand how they are flexible and solid at the same time.

In 1928, W. I. Thomas introduced a maxim that has since become a staple of introductory sociology courses: "if [humans] define situations as real, they are real in their consequences." Often described as the Thomas theorem, it is a simple yet profound insight into the power of social groups to effectively create their own reality. Shared "definitions of the situation" among human actors provide a basis for the most fundamental interactions, and when we do not share these common definitions, it can be hard to see things in the same way. Indeed, it is difficult to imagine collective social life without shared understandings like these, which allow us to take for granted many assumptions about how others in our culture see the world. At the same time, shared definitions of the situation can be oppressive when they marginalize people or groups on the basis of those assumptions. Racism is an example of a shared definition used to support inequalities, constructed from a false view of biological difference.

A definition of the situation has immense consequences if it becomes institutionalized. Institutionalization is the process by which our shared definitions become embedded in structures such as laws, belief systems, and daily practices. While infrastructures are made of material stuff, they are also supported and embedded within a set of institutions that help us interact with them. Tracing the story of the I-35W bridge collapse very quickly connects that specific structure to a diverse set of economic, cultural, and political institutions that created the bridge and responded to its failure. In *The Social Construction of Reality*, Peter Berger and Thomas Luckmann describe the power of institutions to shape our reality and how, despite their basis in human definitions and work, institutions come to have a kind of "objectivity" that make them seem to exist outside human activity: "institutions . . . confront the individual as undeniable facts. The institutions are *there* . . . persistent in their reality, whether [the individual] likes it or not."[1]

The institutionalization of infrastructures explains their apparent solidity as much as or even more than their material structure, as our shared definitions and the power of institutions shape our view of infrastructures as taken-for-granted structures of everyday life when working in our favor and sources of frustration and fatalism when they seem beyond our control. When the temperature in an office is controlled by computers beyond an office worker's authority, governed by HVAC algorithms developed by those with different definitions of the situation, it can seem that institutions and infrastructures are forces of their own will, outside an individual's understanding or mastery.

Berger and Luckmann, however, also emphasize that institutions are fundamentally human creations, a reality of our own making: "the objectivity of the institutional world, however massive it may appear to the individual, is a humanly produced, constructed objectivity."[2] Structures are very real, but their reality is one that we can and do shape every day.

---

1 Peter L. Berger and Thomas Luckmann, *The Social Construction of Reality: A Treatise in the Sociology of Knowledge* (Harmondsworth: Penguin, 1971), 57.
2 Berger and Luckmann, *The Social Construction of Reality*, 57.

Al and the other mechanics also use other technical data in this way, particularly data from sensors built into HVAC systems that measure the temperature of rooms, the status of the system, and other indicators, all of which can be accessed via a centralized computer terminal. In some cases, measurements point them toward the materiality of infrastructure systems. For example, after receiving another complaint about a cold room, a technician named Henry went to the computer and found that the room temperature was 61 degrees Fahrenheit, and although the thermostat was set to a higher temperature, the room was not heating up. Henry promptly went to a valve hidden in the ceiling near this room and found that it had somehow been turned off. When he reopened the valve, the HVAC system brought the room back to the expected temperature.[16]

In other cases, where technical data and their intuition did not point straightforwardly to a faulty part, the mechanics diagnosed the human customers themselves as the problem in need of repair. The mechanics believed that weather conditions and the time of day were key factors influencing calls for repair, especially when someone believed that an office was too cold. For example, a cloudy day might make people feel colder, increasing the number of complaints about cold offices.[17] This again highlights the human body as an indicator of infrastructure trouble.

In particular, the prominence of female bodies in these examples, such as the situation where Al uses the flow meter to show a "lady" that air is indeed coming into her office, points to the gendering of repair discourses. In fact, recent research on the history of HVAC engineering shows how heating and cooling profiles were developed based on male physiologies. Because of this, a gendered discourse of normal temperatures was built into the environments where many people live and work, which defined normal temperature in a way that privileged male bodies (see the "Defining and Controlling Thermal Comfort" box in this chapter).[18] This privileging in turn created a situation that gendered HVAC repair work, based on assumptions about whose bodies are most appropriate for sensing and diagnosing the need for repair.

The social boundaries created by repair discourses can also be used to establish solidarity and community among repair workers. In his book *Talking about Machines*, Julian Orr describes his research with a group of Xerox photocopier repair technicians.[19] Of particular interest are the "war stories" that the (largely male) technicians trade with each other, describing particularly troublesome or notorious repair situations.[20] These stories enable technicians to assert control over an unpredictable job that frequently involves negotiation and improvisation and to establish their masculine identities in opposition to management and customers. The stories serve more practical purposes as well, including the socialization of new technicians and the dissemination of new knowledge and tricks of the trade throughout the repair community. Orr emphasizes that these narratives are no trivial matter, but rather an essential part of getting the work done.[21] They help Xerox manage not only the material functionality of its products, but also the discursive frames that workers and customers use to understand those products and their repair. We return to this theme in more detail in coming chapters, but it is important to emphasize here the value that repair creates for organizations like Xerox. Repair workers not only maintain their products; the commitment to this work through war stories and identification with a set of working discourses makes repair more than just a job. Ultimately this discursive element of repair creates organizational value and corporate profit because workers see themselves as part of a heroic effort to keep systems in good working order.

Repair as Care: The Hot Airplane Problem

Imagine an airplane stuck on the runway, waiting for clearance from air traffic control before takeoff, in the midst of a busy summer holiday with many travelers. The plane waits in a long line of flights ready to depart, and the hot midday sun heats up the plane's cabin. A business traveler in first class presses a call button to summon a flight attendant and complains about feeling too warm. Air travel is one of our most complex infrastructures, merging many technical systems and institutions. At the moment of this interaction, however, the flight attendant is the point of repair for the overheated traveler. This scenario flips many of the features of the cold office problem: hot instead of cold, a moving space where both the air traveler and the flight attendant are only temporary residents, and, based on the fact that women make up a disproportionate number of workers in the service sector, including flight attendants, the likely gender identity of the person providing repair assistance.[22]

The flight attendant might try to repair the traveler's discomfort in a material sense by helping adjust the vent above their seat, providing a cool drink, or talking to the pilot about changing the temperature settings for the cabin. She would also be expected to perform these actions with a smile and overall emotional affect meant to assuage the traveler and make them feel pampered and content. Just as in the cold office situation, this emotional work is integral to successful repair of the infrastructural problem. In sum, the flight attendant is required not only to resolve the material problem, but to do so in a way that shows that she really cares for the over-heated passenger.

The hot airplane problem illustrates the *care work* that a wide range of service workers provide as part of their jobs. Sociologist Arlie Hochschild, in her book *The Managed Heart*, describes this work as "emotional labor," or the management of emotional display through face, gesture, and expression to convey a public demonstration of care.[23] Of course, many people engage in some kind of care work as part of their domestic and social relationships, but Hochschild emphasizes that performing care work as a condition of employment requires the "transmutation" of what we typically think of as private behaviors, performed in the context of close personal relationships, into work skills that are sold in exchange for a wage.[24] This has the effect of decoupling a worker's internal emotional state from that person's public affect. Even flight attendants who are irritated by the passenger or fatigued from working a long shift are expected to put on an apparently genuine "face" of care when promising to help the passenger.[25] While engineers and mechanics aren't typically expected to display the overtly caring affect of a flight attendant or nurse, the cases discussed in this chapter suggest that emotional labor is an important, if often invisible, aspect of their repair practices as well. In the cold office problem, we saw how the work of Al and his colleagues was often as much about shaping the perceptions and feelings of customers as it was about tools and spare parts. This emotional labor is partly disguised by the fact that HVAC mechanics frequently use measurement tools and other technologies to shape the perceptions of customers; they also more often make use of an attitude of professional authority, rather than overt caring, to influence repair outcomes.

As Hochschild notes, however, emotional labor need not be associated with positive affect. Debt collection agents and parole officers, for example, are frequently required to engage in less positive emotional interactions with the people subject to their work. As we saw in the case of the cold office problem, negotiation between the repair person and the office worker involved a kind of back-and-forth discussion and perhaps even debate about the source of the trouble in question. A skilled repair person knows how to seamlessly integrate these affective interactions into an overall flow of repair work that also includes careful use of their body and the tools and technologies at hand. Ultimately this way of seeing the world, as fragile and continually subject to the need for repair, is the basis for an "ethic of care," as described in the writing of María Puig de la Bellacasa and other feminist scholars.[26] An ethic of care is often associated with the so-called caring professions such as nursing, primary education, and social services, domains of work that are traditionally gendered female in many cultures. As we have shown, however, even though mechanically focused repair work is not typically included among these professions, it does in fact require significant emotional labor and an ethic of care, but in this case focused more on things than people per se. The caring professions, too, are commonly tasked with troubleshooting breakdowns of social or material order, like illness or unemployment, further emphasizing their continuity with repair work, including the frequent association of repair with "dirty work." Care work often involves direct work with bodies, chemicals, and other materials that may make the job literally messy, but also brings workers into complex situations where social and material order threaten to break down.[27]

Willie, the rural mechanic, is a good example of a repair worker with a strong ethic of care for things. Harper's ethnography emphasizes Willie's work and how it matters within his local context, where the denizens of Willie's community depend on him to fix their cars and tractors in a cost-effective manner. In so doing, Willie both cares for the community's stock of machinery and plays a crucial role in enabling his customers and their families to maintain financial stability, given that they largely cannot afford to have their machines fixed at standard commercial rates. As a master bricoleur, Willie gains status in his community and a strong sense of his own identity as a skilled, caring repair worker and authority figure among his peers.

The notion that repair work is a kind of caring profession, akin to certain kinds of medical or social work, adds another layer to our understanding of the relationship between human bodies and infrastructures. For example, like HVAC mechanics, nurses perform technical services within a broader context of ensuring the well-being of people who are caught up in social structures that are beyond their control—in this case, the medical system, which is also a sort of infrastructure. In both professions, many workers observe an ethic of care that involves seeing the world as a fragile place in need of constant monitoring and intervening in a compassionate manner to avert catastrophe or just human discomfort.[28] The commonalities between these professions again emphasize the way even our largest infrastructures and social institutions are ultimately grounded in human and material relationships at the local level, which are connected through the medium of the human body and the skills, knowledge, sensations, and emotions that reside there.

The Fragility of People and Things

A fragile item is not necessarily in need of repair, but nevertheless sits in a delicate state of existence, on the edge of stable and broken. Fragility calls our attention to the precarious state of material conditions, including the state of our own material bodies. A cautionary statement reading "FRAGILE" asks us to be care-ful, pointing to the connection between material states and our emotional orientations toward them.[1]

In this chapter, we have emphasized the negotiated character of repair, where any sense of order that we experience in our infrastructural lives comes through the push and pull of people and things, mixed up in settings where they work out the contours of order and disorder. The fragility of these orders is highlighted in the work of STS scholars who question views of technology that emphasize settled states and fixed forms.[2] It seems strange to speak of an unstable stability, and yet that is the state of much materiality.

The fragility of infrastructures is no surprise to the repair workers who spend their days and years maintaining them, and Jérôme Denis and David Pontille emphasize this dual ontology through their research on the way-finding infrastructures of the Paris subway system.[3] Signage is an important means of conceptualizing and navigating a complex urban system like a subway. Denis and Pontille locate the stability of signs in specific policies and standards that help to create a kind of organizational structure for signs and how to interpret them; standardized fonts, colors, and shapes provide consistency for both repair workers and subway users. At the same time, the maintenance workers charged with its repair dwell in a context where decay, vandalism, and even theft constantly shape the state of the signage infrastructure. Sometimes this decay is obvious to commuters or tourists making their way through the tunnels and trains, but other signs of fragility are visible only to the trained eyes of the maintenance workers, such as mold growing on a sign in a damp station near the Seine.[4]

Just as we described for the examples of Al and Willie, who used their bodies to diagnose and repair infrastructures, the subway maintenance staff have a networked sense for the fragility of the signs, meaning that their work involves ongoing surveillance of the Parisian underground. Denis and Pontille describe this orientation as the "care of things," a set of material and improvisatory practices that sees subtle degrees of brokenness in a particular local setting. Caring for things means always seeing them in a vulnerable condition, assessing and reassessing the need for repair.[5] Fragility calls for care, and care points back to the important emotional component inherent in the repair of sociotechnical systems.

1 Benjamin Sims, "Safe Science: Material and Social Order in Laboratory Work," *Social Studies of Science* 35, no. 3 (2005): 333–366.

2 Marianne de Laet and Annemarie Mol, "The Zimbabwe Bush Pump: Mechanics of a Fluid Technology," *Social Studies of Science* 30, no. 2 (2000); Annemarie Mol, *The Logic of Care: Health and the Problem of Patient Choice* (New York: Routledge, 2008); John Law, "The Materials of STS," in *The Oxford Handbook of Material Culture Studies* ed. Dan Hicks and Mary C. Beaudry (New York: Oxford University Press, 2010), 173–188; Jérôme Denis and David Pontille, "Maintenance Work and the Performativity of Urban Inscriptions: The Case of Paris Subway Signs," *Environment and Planning D: Society and Space* 32, no. 3 (2014): 404–416; Jérôme Denis and David Pontille, "Material Ordering and the Care of Things," *Science, Technology, and Human Values* 40, no. 3 (2015): 338–367.

3 Denis and Pontille, "Maintenance Work and the Performativity of Urban Inscriptions"; Denis and Pontille, "Material Ordering and the Care of Things."

4 Denis and Pontille, "Material Ordering and the Care of Things," 348–349.

5 Denis and Pontille, "Material Ordering and the Care of Things," 348–349; Annemarie Mol, Ingunn Moser, and Jeanette Pols, "Care: Putting Practice into Theory," in *Care in Practice: On Tinkering in Clinics, Homes, and Farms*, ed. Annemarie Mol, Ingunn Moser, and Jeanette Pols (Bielefeld, Germany: Transcript Verlag, 2010), 7–26; María Puig de la Bellacasa, "Matters of Care in Technoscience: Assembling Neglected Things," *Social Studies of Science* 41, no. 1 (2011): 85–106; María Puig de la Bellacasa, "'Nothing Comes without Its World': Thinking with Care," *Sociological Review* 60, no. 2 (2012): 197–216; Steven J. Jackson, "Rethinking Repair," in *Media Technologies: Essays on Communication, Materiality and Society*, ed. Tarleton Gillespie, Pablo Boczkowski, and Kirsten Foot (Cambridge, MA: MIT Press, 2014), 221–239; Lara Houston and Steven J. Jackson, "Caring for the 'Next Billion' Mobile Handsets: Opening Proprietary Closures through the Work of Repair," in *Proceedings of the Eighth International Conference on Information and Communication Technologies and Development* (New York: ACM, 2016), 10:1–10:11.

Conclusion: Local Communities of Repair

In this chapter, we have emphasized the role of discourse and negotiation in the work of repair, underscoring the ways repair creates the conditions for local infrastructures to remain relatively stable and invisible elements of our complex sociotechnical lives. Perhaps somewhat ironically, it takes a lot of local work to make something as ubiquitous as infrastructure invisible. More specifically, we have shown how infrastructure repair at a local level is not only accomplished through personal interactions and verbal negotiations between individuals, but is also closely tied to people's material bodies and the skills and sensations embodied within them. These bodies and their perceptions are embedded within local networks of people and things, which enable them to sense potential infrastructural breakdowns and provide access to the collective knowledge and resources required to repair them. Local networks are also the medium for repair work to serve as care work, maintaining and supporting the interpersonal and material elements and relationships that make up our sociotechnical systems. These local interactions are also the genesis of inequality in access to, and control over, infrastructure systems, as some bodies and perceptions are included in local networks and discourses and others become marginalized or even invisible. Indeed, these phenomena are all wrapped up together in some situations, where caring for people and things may actually lead to the invisibility and marginalization of repair workers.

The body, then, is the crossroads where the material and discursive elements of infrastructure and repair intersect. Part of the power of infrastructure lies in its role as a medium that both connects the human body to and buffers it from forces that can operate on a global, geophysical scale.[29] In this chapter, we have provided a detailed account of some of the specific, localized practices surrounding repair work that make this material connection between individual human bodies and global forces both possible and sustainable. To conclude, we explore how these connections can provide a context for community to develop around the practice of repair—in some cases, linking local sites for repair to broader, even global communities. The mediating technologies that increasingly connect our local bodies to global communications networks include consumer products such as cell phones and computers; indeed, technologies such as smartphones are more and more becoming extensions of our bodily senses, memories, and knowledge.[30] But those same products break down—sometimes by design—and call into question the role of consumerism and corporate design methods in shaping these technologies' places in our lives.[31] Research by repair scholars demonstrates these local-global dynamics through two key examples: repair collectives focused on supporting and empowering the users of broken technology to fix it themselves and cell phone repair technicians who draw on diverse resources to work on the devices that have become so integrated in human lives in many parts of the world.

Part of a broader "maker" movement fusing craft, art, and an ethos of sustainable reuse and recycling, repair collectives have emerged, especially in urban settings, at sites around the world. These collectives are often based around workshops where members work on their own projects as well as support those who come seeking assistance to repair a broken household item or to use a tool necessary to complete their own projects.[32] These collectives are an important site for the creation of shared meanings and identities centered around repair. Just as the photocopier repair technicians spent considerable time "talking about machines" with their fellow workers, so collaborative repair settings for amateur fixers create a context for common ground, centered around material objects and their places within our daily rhythms. At the same time, fieldwork in these sites also reveals the ever-present influence of the same gender dynamics that we have described, where the relative value surrounding the work of male and female participants, their choice of projects, and the significance assigned to them are shaped through the gender politics of repair. Lara Houston, Daniela Rosner, and their colleagues describe a repair collective where many of the core participants are white men who have retired from the engineering profession, and in some cases dismiss or denigrate the projects that female attendees of the collective bring to them—this despite the avowed goal of the collective to support the transmission of repair skills and knowledge to new users, inviting new participants to actively engage in the hands-on process of repair.[33] In this way, the practice of repair is policed, and those seeking help may unwittingly "[confront] cultures of masculinity—and associated questions of visibility—developing in and around repair."[34]

As technologies such as internet chat rooms and streaming video services like YouTube allow those with repair needs and interests to communicate translocally, a kind of repair movement is emerging, struggling against the power of manufacturers to prevent consumers from repairing their own products. This "right to repair" movement includes seemingly unlikely bed-fellows, such as farmers hacking into the computerized control systems on their high-tech tractors and websites, such as iFixit.com, showing owners of a smartphone with a broken screen how to repair it themselves (and selling them the parts to do it).[35]

While these examples are focused more on the US context and consum-ers in industrialized nations, cell phones are among the fastest-growing consumer technologies in developing contexts, and their increasing ubiq-uity means that users must also maintain and repair their investment in the technology, though in some cases with fewer resources to discard broken or outdated units and upgrade to the newest models. Houston has conducted field research with repair technicians in Kampala, Uganda, to understand both the repair practices central to those who work on and maintain cell phones for users there, as well as their relationships with other repair prac-titioners both locally and via the internet.[36] These technicians are in some cases independent shop owners who do their work outside of the oversight and support of the major manufacturers, while other repair workers are attached to corporate retail outlets. Repair technicians who work for the major brands have access to tools and infrastructure that connect them with the manufacturers from afar; in some cases, these manufacturers even send their own representatives to set up the repair shops and put in place their own (proprietary) systems for phone repair.[37] Repair technicians who work in independent shops, however, have to rely on their own local, infor-mal networks to share skills, tools, and practices when facing specific repair challenges. In some cases this involves consulting "peers located within walking distance" and, more distantly, "trans-local sites of repair knowl-edge online" via chat rooms and file-sharing websites.[38] This creates a com-munity of repair practitioners forming its own "infrastructures of repair . . . actively pieced together by technicians in the work of searching, connect-ing and collaborating" to find effective fixes.[39] This community grows from the specific market for phones in Uganda, where many of the cell phones that independent technicians repair are recycled units that are imported from industrialized nations and have a second life with users in locations where the market for these phones helps to satisfy demand for cheaper models.

Hacking into the software on the phone is required to open up the phones for use on Ugandan networks, and the repair technicians can find the tools they need through online file-sharing services and creative engineering of specialized tools that allow reprogramming of the phones. Limits on the formation of these communities of repair workers and entrepreneurs, however, come from two sources: competition among rival repair groups promoting the use of their own hacks and products, as well as from limitations imposed by the manufacturers themselves. Similar to the case of an American farmer struggling to fix their own tractor, however, the independent technicians in Kampala are also subject to systems of control implemented by the manufacturers, such as SIM card locks that prevent technicians from accessing and repairing phones with these systems.[40]

In the cases of repair collectives and communities of cell phone repair technicians in developing contexts, we see examples of the potential for repair to create new networks and identities around the technologies and infrastructures for communication enabled by a complex mix of consumer electronics and global networks of exchange. These local communities have the potential to influence broader values and even new institutions. Sociologist Gary Alan Fine argues that "tiny publics" such as local associations and networks make a difference at a larger social scale because their ways of doing things bubble up into collective identities and beliefs that shape larger social structures and forms of order.[41] The right-to-repair movement is one such example of how repair might bring together coalitions of interests to influence cultural and legal change at a national or even international level. Given the scale of large infrastructural systems, their scope can potentially dissuade attempts to change their structure and politics, but the way that users engage locally with technology provides a potential site for intervention and change. At the same time, the examples in this section point to the continued power of corporate interests in shaping our uses—including the preclusion of unwanted uses—as well as potential conflict among users with competing interests and identities invested in repair. Local bodies and materiality connect with these diverse infrastructural elements through time and space. The coming chapters continue to explore how repair is negotiated at scales that link the local to regional, national, and global networks.

3   Bridging Scales: The Local Negotiation of Systemic Repair

In the beginning, [the Coronado Bridge] was a symbol of devastation. A lot of people feel it ruined Barrio Logan. When it first opened, it had a terrible, terrific impact. But now it's a piece of living art. It has a melodic effect on my life, my vision. It's a strange thing to say, but I love the bridge.[1]

—Salvador Torres, muralist and activist, Barrio Logan, 1989

Three Episodes of Repair in the Life of a Bridge

This chapter tells the story of the San Diego–Coronado Bridge, a sweeping, iconic structure that connects the City of San Diego with the much smaller City of Coronado across San Diego Bay. We first explore the origins of the bridge in the 1950s and 1960s, and how the City of Coronado, the City of San Diego, the US Navy, the business community, and the State of California came together to create a network of associations needed to get the bridge approved and built. This is also the story of how the community that would be affected the most by the bridge—a largely Mexican American neighborhood of San Diego called Barrio Logan—became a path of least resistance for planners and engineers. We then dig into three episodes of repair during the lifetime of the bridge, each of which transformed the power structures and networks of relationships surrounding the bridge, or other bridges across California, in a different way. Taken together, these repair episodes show that even a seemingly unchanging piece of infrastructure like the San Diego–Coronado Bridge is in fact a living artifact, the focal point of sociotechnical networks that change and adapt over its lifetime, giving it new purposes and meanings over time.

In the first repair episode, in the 1970s, the community of Barrio Logan and its allies used a combination of protest, artistic transformation, control over land, cultural resources, and political action to disrupt and repair the network of associations around the bridge in a way that gave them a more prominent place in it. The result was the development of Chicano Park under the bridge approach ramps and a series of monumental murals painted on the columns in the park, which have since become nationally recognized as a singular artistic achievement and emblem of the Chicano civil rights movement.[2] This case study emphasizes the possibility of repair from below, when actors who have been marginalized by a sociotechnical system use repair to transform that system in ways that advance their interests.

The second repair episode, spanning the 1970s through the 1990s, breaks away from the Coronado Bridge (as it is known locally) to examine the larger context of bridge engineering in California. In this episode, California Department of Transportation (Caltrans) bridge engineers came to understand that their design practices did not adequately protect bridges from earthquake damage. We follow their efforts to transform their design practices through new ways of observing earthquake damage and new relationships with the research community, and their efforts to retrofit thousands of bridges throughout the state to meet these new standards. This episode highlights how infrastructure systems can come to be seen as broken even when they are physically unchanged and describes how experts may seek to repair a sociotechnical system they are part of in order to maintain the integrity of the system and their own status within it.

In the final repair episode, these threads come together through the story of the seismic retrofit of the Coronado Bridge. We show how the retrofit project forced experts and a more empowered Barrio Logan community to confront each other over their very different views of the bridge. The subtle series of displacements and negotiations that followed redrew the network of associations around the bridge yet again, allowing retrofit to go forward and Chicano Park and its murals to be preserved. This episode emphasizes that repair at this scale can be a tricky process, balancing myriad local and systemic factors to achieve a broadly acceptable outcome.

From Local to Systemic Repair

In chapter 2, we focused on the negotiation of repair at small scales, emphasizing the immediate, local engagements of humans with their material and social environments. In this chapter, we shift the perspective to a scale where local negotiations and engagements over infrastructure repair are still important, but participants are also engaged in deliberate efforts to force and manage change at a systemic level. This gives us an opportunity to observe how local and systemic aspects of infrastructural repair interact and how people manage the tensions between scales.

As we have discussed, local negotiations over repair are deeply connected to the material and social configurations that define the specific settings where they take place. This corresponds to the bottom-up analytical perspective on infrastructure, with its focus on how infrastructures enable activities in specific places. Repair workers must have a deep understanding of the local elements of infrastructure systems and how they connect with local material and social arrangements in order to do their jobs, and indeed all infrastructure users develop a similar sense of how to make things work locally. We have discussed this way of knowing about infrastructure in terms of embodiment—the idea that the work of repair involves skillfully deploying the human body and its senses in relation to a local setting—and in terms of the networked body, which emphasizes that the skill of repair work is embedded within a local network of people and things as much as it is in an individual's body.

To understand the systemic aspects of repair, we have to expand this view by looking at networks of people and things that transcend localized settings. We use the term *network* here as it is widely used in the field of STS, referring not only to physically realized webs like road networks but also to the wider set of connections among people, things, and concepts that we can trace outside particular sites like the Coronado Bridge or a cold office.[3] *Network* is a useful metaphor because it provides a way of describing the complexity of sociotechnical systems, as well as the idea that they have no firm boundaries, but instead ultimately connect to form a global web of relationships. This gives us a way of describing the push and pull between human and material agency not just on a local level, but also at the larger scales needed to understand whole infrastructures and how they are integrated into larger social and cultural settings and structures of power and control.

From this broader network perspective, our emphasis now shifts from embodied sensations and abilities toward the more abstract realm of human abilities involved in designing and controlling large sociotechnical systems; from locally embedded skills and knowledge toward systematized bodies of science and engineering knowledge; from relationships among individuals toward relationships among communities, organizations, and professions; from interpersonal negotiations of power toward the deep connections between infrastructure systems and larger social hierarchies and power structures; and from local tensions toward public political controversies. This is a matter of emphasis, however; how the local, relational aspects of repair remain important even as we move to these larger scales is a central theme of this chapter.

Although we have already described how infrastructural repair intersects with differentials in power among social groups, in this chapter we consider in more detail how the material configuration of infrastructures, at a systemic level, both reflects preexisting power structures and serves to reinforce them, either unintentionally or by design. This happens because the interests of powerful groups tend to be addressed in infrastructure design, while the interests of the less powerful are often excluded or simply remain invisible to system designers. Infrastructure systems are able to reinforce or extend the interests of those in power by casting their interests in material form. This does not mean that technological structures directly reflect political arrangements, as philosopher Langdon Winner has argued.[4] The point is that infrastructures, as sociotechnical systems, are thoroughly integrated with social and political structures through their design and use. These sociotechnical arrangements are not, however, inherently stable over time; they are subject to repair, maintenance, and renegotiation, creating opportunities for political change that can undermine as well as reinforce powerful interests.

The influence of power differentials on infrastructure is often reflected in unequal distributions of costs and benefits. Benefits such as faster transportation, easier access to energy, or broadband internet connections may end up being more accessible to communities that are influential in infrastructure planning, while others are bypassed by internet services, stuck with polluting power plants, and faced with demolition of housing to make way for highways. Infrastructure development may result in economic disruption, health impacts, and lack of services for those communities, which may further influence their access to social and economic power. Coupled with other legal and cultural means of reinforcing racial and economic divisions, in cities this has led to large-scale patchworks of neighborhoods, with stark geographical divisions between haves and have-nots materially reinforced by infrastructural arrangements.[5]

Despite these effects, the tension between local and systemic aspects of repair can sometimes make infrastructure a problematic ally for advancing the interests of the powerful. Design and management of infrastructure systems is often coordinated at what Bruno Latour has called *centers of calculation*: places where data, information, and knowledge collected from across a wider territory are concentrated and made accessible to manipulation and analysis.[6] Without these centers, many modern infrastructures could not easily function: air traffic control centers keep the air transportation system in order; technical innovations and decisions made in Silicon Valley maintain the structure of services that defines the modern internet; and bridges and road systems are planned and designed in administrative centers and capital cities. These are crucial centers of control over infrastructure operations and planning; powerful interests often have access to or influence over them, while more peripheral or marginalized groups usually do not. At the same time, however, the power of these centers is entirely dependent on their relationships with the many different peripheral locations where data are collected and local embodiments of infrastructure systems are built and used. To design and maintain a road system, for example, somebody has to measure local traffic patterns, test soil conditions, survey the local landscape, hire construction workers, understand local overlaps between infrastructure systems, and manage political and public concerns. The system as a whole cannot function without understanding and managing these local relational aspects of infrastructure.

The tension between local, relational aspects of infrastructure and centralized, systemic aspects is part of what makes infrastructure systems vulnerable to disruption. For example, because infrastructures like roads and power lines are often so closely connected to the landscapes they pass through, they can be vulnerable to the local effects of disasters like hurricanes and earthquakes. Infrastructure systems are also locally vulnerable to sabotage and misuse. A computer virus, for example, can be created and launched from just about any computer connected to the internet yet have widespread global impacts. As we explore in this chapter, breakdown and repair of infrastructure systems is often prominent at these intersections between the local and global aspects of infrastructure; think about the desperate efforts to restore infrastructure services to New Orleans following Hurricane Katrina in 2005 or the computer antivirus industry, which provides tools for tracking and repairing damage from an endless variety of attacks that threaten computers and networking infrastructure. More important for this chapter, the local embedding of infrastructure not only makes it vulnerable to disaster or sabotage; it can also provide points of leverage for groups that live with the negative effects of infrastructure to gain access to the power structures that govern those systems.

These topics tie together several of the elements of our tool kit for understanding repair. They show how the materiality of infrastructure systems reinforces relationships of power and invisibility. These relationships are shaped by scale—in particular, the way infrastructure systems inherently link local settings and relationships to large-scale technological and social structures. In addition, we touch on the relationship between materiality and discourse by examining how discourses of slippage and repair emerge out of engagements with the material world and how they can then be translated back into material forms. This will help us understand the differences between repair as maintenance and repair as transformation.

Building the Bridge

Coronado is both a community and geographic entity that forms part of the boundary of San Diego Bay, across the bay from the City of San Diego. Although sometimes referred to as an island, it is connected to the mainland far south of downtown San Diego by a narrow, sandy strand. The City of Coronado is immediately adjacent to the Naval Air Station located on North Island, which was a separate piece of land until the channel separating it from Coronado was filled in. Coronado was originally developed as a beach resort community and is still home to the massive, architecturally significant Hotel Del Coronado.[7] Today, Coronado is primarily residential, and is one of the wealthiest communities in the San Diego area.

There had been plans for a bridge between San Diego and Coronado as far back as the 1920s, supported mainly by the tourist-oriented business community. These plans had all failed due to intense opposition from Coronado residents who wanted to preserve its small-town character and isolation from San Diego and Navy concerns about a bridge interfering with the movement of large ships in and out of San Diego Bay. The situation changed in the 1950s and 1960s under the administration of California Governor Edmund G. "Pat" Brown, a powerful advocate for the construction of freeways and bridges with strong political connections in Washington,

Figure 3.1
The San Diego–Coronado Bridge, with downtown San Diego in the center of the image, Coronado to the left, and Barrio Logan at the far right. Photograph by Frank McKenna, 2016. https://commons.wikimedia.org/wiki/File:San_Diego-Coronado_Bridge _by_Frank_Mckenna.jpg. Published under Creative Commons CC0 1.0 Public Domain Dedication.

DC. Design for the current bridge began in the early 1960s, and the bridge was completed in 1969. Figure 3.1 provides an aerial view of the bridge.[8]

In an interview, one of the (San Francisco–based) bridge engineers recalled that the idea of a bridge was quite controversial in both San Diego and Coronado; he said he was always "greeted very coldly" by local officials, and a city council member told him to "take your bridge and go right back to San Francisco."[9] The business community was more supportive; another interviewee described the bridge as a "political payoff" by Governor Brown to major landowners in Coronado who had supported his campaign. This interviewee also described how the bridge designers worked to mitigate local concerns by bringing in a prominent San Diego architect to work as a consultant on aesthetic details, including design of the gracefully arched bridge supports and the decision to paint the metal portion of the bridge blue instead of red.[10]

Much of the bridge's layout ended up being dictated by Navy concerns. In particular, the Navy insisted on a clearance of at least two hundred feet over the main shipping channel, which would have made a straight bridge between San Diego and Coronado dangerously steep. The bridge designers addressed this concern by adding a sweeping, 90 degree curve to the bridge as it crosses the bay, making the bridge long enough to accommodate the extra height while maintaining a reasonable grade for the roadway. The location of the two ends of the bridge was dictated in part by the need to place the bridge's highest point directly over the main shipping lane, which is closer to the San Diego side of the bay. As a result, the bridge is much lower on the Coronado side, ending at the shoreline, while it remains quite tall for a considerable distance inland on the San Diego side, until it connects to Interstate 5.

This configuration was made possible in part because the San Diego end of the bridge ran through the already disrupted Mexican American neighborhood of Barrio Logan, which had low property values and little political influence. In an interview, one of the engineers of the bridge remembered some opposition from the neighborhood, but given financial constraints and other local opposition, he recalled that bridge planners saw this configuration as "the path of least resistance" at the time.[11] Minimizing the visible impact of the bridge on the Coronado side may have helped soften the opposition from that wealthier, more politically connected community.

An Infrastructural Inversion
(Benjamin Sims)

As a graduate student in San Diego in the 1990s, my initial experience of the Coronado Bridge is literally top-down: as a driver, occasionally using the bridge to access businesses in Coronado, and more often observing the sweeping curve of its deck across the bay from a distance on the highway or when flying into the airport. I have not considered that there might be anything interesting going on *under* the bridge, except maybe the occasional passing of a big Navy ship. The neighborhoods around and under the bridge approaches on the San Diego side are places I simply bypass without thinking much about them.

My perspective changes on a bright November day in 1996 when I join a group of UCSD engineering graduate students to tour the bridge and learn about its seismic retrofit. By this time, I know the name of the San Diego neighborhood it passes over—Barrio Logan—and the park under the bridge approaches—Chicano Park—and that there are murals painted on some of the columns of the approach ramps, but not much beyond that. As I approach the park in a carful of aspiring bridge engineers, the first mural we see says "No Retrofitting" (figure 3.2). The engineers laugh, slightly nervously, I think.

The murals make a striking first impression on exiting the car. These aren't just a few discrete murals painted on the side of the bridge columns, as I had imagined. Rather, they are huge, intensely colored works of art, in many cases wrapping all the way around the columns and up into the bridge deck supports; some are thirty or forty feet tall, and there are dozens of them, concentrated around a pavilion in the center of the park modeled after an Aztec temple. They transform the space under the approach ramps from a gray concrete wasteland into something resembling an outdoor cathedral, with the murals standing in for stained glass windows. This impression is intensified by the arched bridge columns marching out across the bay into the distance. A man plays catch with a boy; a woman reads a book at a table; a park worker picks up trash; people rest in the sun; a pickup truck drives by, its occupants looking at our group curiously. We hear about the retrofit project, the murals, and the community from a Caltrans community liaison and about the technical aspects of the retrofit from a UCSD professor. We then get back in our cars and head across the top of the bridge to the park at the other end, in the wealthy resort city of Coronado. The contrast is striking. This park resembles a tropical beach, with palm trees swaying in the breeze. The bridge is barely present, as its structure touches down right at the waterline, and the park is largely empty of people.

Caltrans workers then unlock a gate through a chain-link fence, and we climb onto a catwalk under the bridge. As we walk along the mesh-floored catwalk, we get higher and higher, dizzyingly high above the bay. On our walk, we stop periodically to check the hinges at the top of some of the columns; the engineers take some measurements here and there. Eventually we reach a ladder that takes us up into the inside of the central bridge structure, a single cavernous, hollow box beam that supports the entire roadway above. Bright lights pierce the darkness as we proceed along a catwalk on one wall, occasionally catching a glimpse through a portal at the water far below. Finally we exit the beam and climb down a slightly panic-inducing ladder on the *outside* of the beam and onto another mesh walkway encircling the top of one of the central, tallest columns of the bridge. The view is stunning, taking in downtown San Diego, the bay, Coronado, and several large Navy facilities. My feet are sore and my arms are tired from climbing. The bridge seems immense from below; there is simply so much more of it than is apparent from above. And there are whole communities—among them bridge engineers, maintenance workers, muralists, and Barrio Logan residents—that experience the bridge primarily from this side. Infrastructural inversion (see chapter 1) is not just a methodology for these communities; it's a lived experience.

Figure 3.2
Chicano Park, with *No Retrofitting* mural (1995) in the foreground and *We Saved the Murals* (1997) mural partially obscured on the wall behind it. Photograph by Benjamin Sims, 2018.

Repairing Barrio Logan

Barrio Logan is part of the larger San Diego neighborhood of Logan Heights, located south of the city's downtown. Logan Heights was historically one of the city's most diverse neighborhoods, known as a center of Mexican American, African American, Asian American, and immigrant populations through at least the 1960s.[12] The area of the neighborhood closest to the waterfront, in particular, had become a largely Mexican American community by the 1940s, and at some point it became known as Barrio Logan.[13]

This history connects to an overall process of "barrioization" that unfolded in cities throughout California and the southwestern United States in the first half of the twentieth century.[14] Mexican American populations, which had already been marginalized and displaced by the enormous wave of Anglo settlement that began in the nineteenth century, were increasingly segregated into specific neighborhoods, or barrios. This process was driven by racially restrictive covenants, as well as periodic episodes of slum clearance and urban renewal that targeted existing Mexican American neighborhoods. Although this process was driven by forces external to these communities, the barrios also became cultural centers that fostered consciousness of a common identity and a sense of connection to a particular place within the urban fabric.[15]

Once established, however, the barrios were again often targeted for disruptive and polluting industrial and infrastructural development, while suffering from a lack of city investment in infrastructure for residents.[16] In particular, development of urban interstate highways in California during the 1950s and 1960s often had disproportionate impacts on Mexican American neighborhoods, part of a general trend where poor and minority neighborhoods were treated as a path of least resistance for urban highway development throughout the United States. By the mid-1960s, however, "freeway revolts" were breaking out in many cities, and residents of affected communities, including barrios throughout California, became increasingly assertive in protesting and otherwise opposing unwanted infrastructure development.[17]

Barrio Logan exemplifies many of these patterns. During World War II, the neighborhood was cut off from the waterfront by defense-related industrial development along the bay. In the 1950s, the city changed the zoning of the neighborhood from residential to mixed use, resulting in displacement of housing by junkyards and industrial plants. The neighborhood was then partially demolished and cut off from the rest of Logan Heights by the construction of Interstate 5 in the early 1960s. Finally, the construction of the Coronado Bridge bisected what was left of the neighborhood, resulting in more destruction and covering a portion of it with elevated ramps connecting the bridge to the Interstate. Due to these disruptions, the population of the Barrio Logan area fell from around twenty thousand prior to World War II to about five thousand by 1979.[18] These developments had a profound and negative impact on the community. US poet laureate Juan Felipe Herrera, who grew up in the neighborhood, later likened the result to "the coming of Godzilla; / a kind of urban ghost town."[19]

By the time the Coronado Bridge was completed in the late 1960s, Barrio Logan and the rest of the Mexican American community in San Diego was experiencing a political awakening, inspired by the growing Chicano civil rights movement, including Cesar Chavez and the United Farm Workers and the Chicano youth movement. During construction of the bridge, Barrio Logan residents had requested that the city build a community park under the approach ramps, and the city had leased state land under the bridge with the apparent intention of doing so. With the expectation of a park, the community was shocked when, one morning in April 1970, bulldozers appeared under the bridge—to begin building not a park but a new San Diego headquarters for the California Highway Patrol. The news quickly spread throughout the neighborhood and the larger community. Convinced this would complete the destruction of the neighborhood, residents quickly mobilized, with politically active young people taking a leading role. A large crowd of protesters surrounded the site, forcing a halt to construction. Protesters occupied the site for twelve days, until the city agreed to negotiations on a park. The city ultimately acquired the land slated for the Highway Patrol headquarters and formally agreed to build the park.[20]

Artists played a central role in the vision for the park. Salvador Torres, a resident of Barrio Logan who had studied art in San Francisco and then returned to the neighborhood, took on a primary leadership role. Despite the destruction the bridge had wrought, Torres began to develop a dramatic vision of artistic transformation centering around monumental murals painted on the bridge columns in the park, inspired in part by murals he had recently seen by Diego Rivera, José Clemente Orozco, David Alfaro Siqueiros, and other icons of Mexican muralism during a trip to Mexico.[21] Torres and others in the community began to see aesthetic qualities in the appearance of the bridge as seen from below, with its arched support structures running out into the bay that one resident likened to a cathedral.[22] Meanwhile, the city had been slow to make improvements in the park area. Partly in reaction to this, Torres gathered a group of about a dozen artists, many of whom were affiliated with the Centro Cultural de la Raza in nearby Balboa Park, to develop plans for painting murals. After several years of lobbying, the group was granted permission by the San Diego Coronado Bridge Authority to begin work in 1973.[23]

A large proportion of the approximately ninety murals that now exist in Chicano Park were painted between 1973 and 1981. Early murals were painted primarily by San Diego–based artists, but later artists from elsewhere in California and the Southwest were invited to participate.[24] The murals reflect a range of themes (figures 3.3 and 3.4), including Aztec

Figure 3.3
Chicano Park, with *Coatlicue*, *Chicano Pinto Union*, and *Chicano Park Takeover* murals (all 1978) in the foreground and the *kiosko* or pavilion partially obscured in the center. Photograph by Benjamin Sims, 2018.

imagery, depictions of historical events (including the park takeover itself), political symbols and slogans, depictions of significant historical figures and artists, and more abstract artistic statements. Examples include a depiction of a heavily muscled man who appears to be holding up the roadway with his hands (figure 3.5) and a strikingly colorful modernist portrait of Rivera, Orozco, Siqueiros and Frida Kahlo ("Los Grandes").[25]

Transformation from the Margins

The Chicano Park murals are a vivid example of transformational repair executed from the margins of an infrastructure system. As the history of barrios throughout California indicates, the distribution of costs and

Figure 3.4
Chicano Park, with *Renacimiento* (*Birth of La Raza*) (1974) and *Niños del Mundo* (1975) murals in the foreground and *La Familia* and *Inlakesh* (a.k.a. *Mandala Mural*) (both 1975) murals behind them. A children's playground is visible in the background. Photograph by Benjamin Sims, 2018.

benefits from infrastructure is often highly unequal, and in many cases, the communities that experience the least benefit from infrastructure construction also experience the most negative impacts. The bridge and freeways built through Barrio Logan, for example, provided little additional transportation connectivity to the already centrally located neighborhood, but had an enormous negative effect on the quality of life of residents. Particularly during the twentieth century in the United States, locally massive infrastructures like freeways and bridges were often built through politically and economically marginalized areas or created marginalized or abandoned spaces within the urban fabric. People who live in these areas may experience infrastructure from a unique perspective. The significance of a

Figure 3.5
Chicano Park, with *Colossus* mural (1975) in the foreground and *Children's Mural* (1973) partially obscured behind it. Part of the Logan Avenue business district is visible in the left background. Photograph by Benjamin Sims, 2019.

bridge for most of its users, for example, may be as an element of connectivity in a larger road network, but to those living under it, it may loom large as an object in itself—a barrier, perhaps, or an aesthetic intrusion. The workers who repair and maintain infrastructure similarly experience it from a different perspective than most users do. These situated experiences are infrastructural inversions as seen by ordinary people in their everyday lives. What these perspectives have in common is that they are often experienced by people who are marginalized in some way—those who in some sense have limited visibility within or control over the sociotechnical arrangements surrounding infrastructure.

What is particularly interesting about the Chicano Park case is that the Barrio Logan community was able to leverage the limited amount of local control it had over the disrupted space occupied by the Coronado Bridge to effect a transformation in the localized meaning of the structure. They did this by executing a comprehensive material transformation of the land and bridge surfaces in their neighborhood. This was explicitly framed in terms of consecrating a symbolically and culturally significant community space, as described by Raúl Homero Villa in his study of the park:

> The founding of the park involved substantial cultural production and ritual activities that were meant to expressively conjure a specific sense of place beyond the construction of a mere physical environment. From the consecration of the ground for the planting of the first greenery, to the design and construction of a central pyramidal *kiosco* (kiosk), to the pictorial embellishments and motifs of the murals painted on the support pillars of the bridge ramps, to the practicing of pre-Columbian *danza* on the site, a spirit of indigenous identification with the land infused the multiform physical embellishments and cultural enactments of the site.[26]

The transformation of the bridge and the surrounding space was closely connected to patterns of power and invisibility surrounding the bridge. The bridge devastated Barrio Logan because planners decided they could ignore the community and its concerns, but by taking local control of the space disrupted by the bridge, the community was able to stake a unique and increasingly visible claim within the sociotechnical networks surrounding the bridge. This became a particularly important point of leverage as new environmental impact and cultural preservation laws came into play during the 1970s and shows that transformative repair need not come from actors who have centralized control over an infrastructure system. Infrastructure systems can create sociotechnical connections between centers of power and control and the less visible local spaces and disenfranchised communities that are often most negatively impacted by those systems. Because of these connections, there is often at least a small opening for infrastructures to escape centralized control, providing populations that are otherwise marginalized with potential points of local leverage to effect change. Thus, infrastructure systems can play a major role in shaping and connecting local power structures with the spatial organization of communities. Processes of repair and maintenance make this an ongoing and dynamic connection, as we explore further in this chapter.

Repair Networks and Slippage

The history of the San Diego–Coronado Bridge highlights the way infrastructural artifacts become embedded in broader networks of political, economic, organizational, and cultural relationships and how repair can transform those networks. We see this in the way the plan of the bridge emerged as a compromise between engineering considerations and powerful national, state, and local interests, with the Barrio Logan community largely excluded from the process—an exclusion rooted in larger patterns of economic inequality, cultural disregard, and racism. But we also see a different network of relations evolving around the bridge, through the park and the murals—one in which an excluded community leveraged what power it possessed to force its way into the negotiation of interests around the bridge. The community then expanded this foothold by making the park and murals a center of political activism in San Diego, as well as an increasingly widely recognized cultural landmark. This is a key example of what we call *repair as transformation*. By a combination of political acts and material transformations, people took an infrastructural space that was not functional for them and turned it into a center of culture and community identity, and ultimately into a source of greater political power. In doing so, artists like Salvador Torres implemented a radically different vision of what a bridge could be, as seen from below, and made it stick. The repair here was both a material transformation of the bridge and a realignment of the sociotechnical networks surrounding it.

When we talk about sociotechnical repair and repair networks on a larger scale, this is often what repair looks like: not a crumbling of something solid that is then restored, but a reconsideration of what functional infrastructure looks like, an articulation of a need for change, and a process of correction. This connects to the idea of *slippage*, which we introduced in the first chapter. Slippage is a discursive framing of repair that begins with identifying some kind of breakdown or inadequacy in a system as it exists, then contrasting that with an ideal state that could be achieved through repair. This can be as simple as a statement like, "It's too cold in this office. Can you fix the thermostat?" or as complex as the scientific research and political debates around global climate change and what ought to be done to create a better outcome. Slippage might be framed in terms of a breakdown in the material form of a system due to damage or aging, for example, or changing needs that the material form of a system can no longer support, making it obsolete. More often, it takes more complex forms, in which changing material conditions, user needs, and technical knowledge might all play a role. As controversies over issues like climate change suggest, people often have very different perspectives and beliefs about whether any slippage has occurred at all, whether it warrants repair, and what a possible repair might look like, all of which have the potential to generate fierce arguments and power struggles.

Barrio Logan activists articulated slippage around the Coronado Bridge primarily in terms of the needs of the community and how infrastructure should accommodate those needs. The design and construction of the bridge did not take those needs into consideration. Although it was too late to make fundamental changes to the material configuration of the bridge, the community saw an opportunity to reimagine the bridge in a way that met some of their needs by transforming the land beneath it into a community space and the surfaces of its columns into works of art. As we will explore in more detail later in this chapter, these acts of repair also reinforced a range of other transformations in the networks surrounding the bridge, including the emergence of more unified community leadership through organizations like the Chicano Park Steering Committee, increased recognition of Barrio Logan's interests among local political leaders and institutions, and use of state and federal historical designations to protect the murals and the park. Taken together, these changes amounted to a significant transformation not only in the material state of the bridge, but also in the relationships between Barrio Logan, the bridge, and local, state, and national power structures. This diversity of interconnected changes in networks is common in larger-scale repair and is why a sociotechnical perspective is needed to understand this process.

These local developments also intersected with broader cultural trends, including increased Chicano activism across California and the United States and increased consideration of the impacts of large infrastructure projects on communities and historical sites. The latter development was implemented through legislation like the National Historic Preservation Act (NHPA) of 1966 and the National Environmental Policy Act (NEPA) of 1969, which requires federal agencies to assess the impact of large infrastructure projects not only on the natural environment, but also on historical and cultural resources and community standards of living. These laws were partially motivated by the negative experiences of many communities with interstate highway construction projects in urban areas during the 1950s and 1960s, which often disrupted the urban fabric without real consideration of the consequences. This provided a stronger legal basis for communities like Barrio Logan to exert control over local infrastructure developments.[27]

Repairing Earthquake-Resistant Design Practice at Caltrans

While the Barrio Logan community was working to address the slippage it had articulated around the Coronado Bridge, this bridge and others around the state became the focus of another discourse of slippage and obsolescence, this one driven by engineers and focusing on the threat to bridges from earthquakes. These two discourses eventually collided in the 1990s when engineers decided the columns of the Coronado Bridge needed to be retrofitted to improve their earthquake resistance, threatening the existence of the murals.

Starting in the 1970s, engineers had developed new understandings about the impact of earthquakes on structures, particularly those made of reinforced concrete. Based on this understanding, as well as unexpected damage seen during earthquakes, the bridge engineering community came to believe that past design practices left many older bridges at risk of collapsing in earthquakes. Here, we focus on how this discourse of obsolescence was articulated and managed by engineers at Caltrans, who are responsible for the design and maintenance of bridges, overpasses, and other roadway structures throughout the state, including the Coronado Bridge.

Prior to the 1960s, Caltrans engineers lacked tools for calculating how a bridge responds dynamically to earthquake stresses based on its particular structural configuration. Because of this, they accounted for earthquake risks in bridge design primarily through large safety margins, resulting in heavily built bridges that were strong but not necessarily very resilient to ground shaking. As engineering methods and computational tools evolved during the 1960s, it became easier to calculate these dynamic responses. This led to changes in design standards that allowed for more slender, economical, and aesthetically interesting structures. However, engineers did not gain much experience with major earthquakes during the rapid expansion of the state's highway network in the 1950s and 1960s, so they had little practical understanding of how their designs might stand up to seismic forces.

This all changed in 1971 when a 6.6 magnitude earthquake struck the San Fernando Valley, then a rapidly developing northern area of Los Angeles. Several brand-new freeway overpasses that had not yet opened to traffic collapsed during the quake, which was a great surprise to Caltrans engineers since they were designed according to the latest earthquake resistance standards. On investigation, engineers discovered two main types of failure. First, the earthquake caused bridge decks to pull apart at expansion joints, allowing segments of the roadway to fall to the ground. Second, within some of the bridge columns, the cages of steel reinforcement ("rebar") that prevent concrete from breaking up under stress had ruptured during the earthquake, causing the columns to disintegrate (figure 3.6).

The contrast between the expected earthquake performance of newly designed bridges and their actual performance led Caltrans engineers to articulate an overall breakdown in their design practices: existing state-of-the-art engineering approaches resulted in bridges that could not stand up to a relatively modest earthquake, so those methods would need to change. While there was early consensus on some required changes, other slippages were understood in general terms, but remained frustratingly difficult to articulate and repair. This led to an approximately twenty-year period of incremental evolution in Caltrans design practice, with a real sense of confidence that design and retrofit methods were truly adequate to protect bridges from earthquakes emerging only in the 1990s.

During the 1970s and 1980s, most of the changes Caltrans made to bridge design standards were at the level of design details—that is, best practices in specific areas, like how to tie rebar together at joints, that are known to improve performance but are not part of an overall design methodology. In response to the observed column failures, many of the changes focused on increasing the amount and continuity of rebar in columns, such as using continuous spirals of rebar within a column instead of individual hoops. To address the joint failures that caused roadway sections to fall, engineers came up with the idea of tying joints together with steel cables or rods. Since this modification could be applied to bridges already built, it raised the prospect of *seismic retrofit* of California bridges for the first time. Joint retrofit work progressed slowly under funding constraints throughout this period. Caltrans had no immediate method for retrofitting columns, however, so that idea languished until the mid-1980s.

Beyond engineering changes, more significant efforts to repair and update Caltrans design practice were afoot. Beginning in the 1970s, Caltrans engineers began to comprehensively rethink how they gained earthquake engineering knowledge and incorporated it into design practice.

Figure 3.6
Column in the Interstate 5/Interstate 210 interchange, heavily damaged by the 1971 San Fernando, California, earthquake, showing rebar failure. Photograph by Edward C. Brinley, 1971. From the Earthquake Engineering Online Archive/NISEE-PEER E-library, https://nisee.berkeley.edu/elibrary/Image/S4385 (membership required for access). Courtesy of NISEE-PEER, University of California, Berkeley.

This articulated a slippage between the more basic, standardized design approaches Caltrans had taken through the 1960s and a desire for a more dynamic, engaged form of practice that actively incorporated outside input.

The first expression of this new design practice was an increased emphasis on investigating the impact of actual earthquakes on Caltrans bridges through the establishment of a formal Post-Earthquake Investigation Team. During the 1980s, Caltrans also developed an increasingly close relationship with the growing earthquake engineering research community. In the mid-1980s, Caltrans began to take a new look at column retrofit methods and was quickly drawn to a method of reinforcing columns by enclosing them in steel jackets, which was associated with a highly respected engineer from New Zealand, Nigel Priestley (figure 3.7). As luck would have it, the University of California, San Diego (UCSD), opened a major structural engineering laboratory in 1986 that was capable of testing bridge columns at nearly full scale and promptly recruited Priestley as a faculty member. Caltrans began to collaborate closely with Priestley and others at UCSD and fund a research program there. This relationship took on a central role in the development of Caltrans engineering knowledge, design practice, and retrofit strategies through the 1990s and beyond.

Public Engagement with Earthquake Risks

Throughout the 1970s and 1980s, Caltrans engineering practice steadily evolved in the face of almost complete public indifference to the earthquake threat to bridges. Although Caltrans engineers saw the 1971 San Fernando earthquake as a major wake-up call, the lack of casualties gave it less significance in the eyes of the public. Unlike Caltrans engineers, the general public never came to perceive existing bridges as obsolete from a seismic perspective. This indifference translated into little political pressure or funding to support seismic retrofit efforts.

Public perceptions changed drastically on October 17, 1989, when the 6.9 magnitude Loma Prieta earthquake struck near San Francisco on the afternoon of a World Series baseball game. The temblor killed sixty-three people, forty-one of them in the collapse of the Cypress Viaduct, a double-decked Caltrans structure that carried a section of Interstate 880 in Oakland.[28] Caltrans bridge engineering practices and the slow pace of seismic retrofit were suddenly a major news story. The story was prominently framed through slippage rhetoric, with journalists, politicians, and members of the general

Figure 3.7
Steel shell retrofit in position around a flared column, ready for grouting and welding, California State Route 52 overpass over Genesee Avenue, San Diego, California. The column footing is also being retrofitted. Photograph by Benjamin Sims, 1998.

public all expressing shock that not all California bridges were able to with-stand this level of shaking and calling for something to be done.

Many of the initial media stories zeroed in on Caltrans engineers as the source of the problem, and for several days, it looked as if they might take the blame. However, as journalists and politicians uncovered Caltrans's existing retrofit efforts, Caltrans officials deftly steered them toward a dif-ferent story, in which Caltrans had made extensive efforts to reform its design practices and retrofit bridges during the 1970s and 1980s, but were unable to proceed with the required urgency due to lack of funding. This theme was summarized concisely by a highway contractor quoted in the *San Francisco Chronicle*: "It's the classic problem—Caltrans wants more money for highway maintenance and you can't get the politicians off the dime."[29] With the threat of political blowback, the governor appointed a board of inquiry made up mainly of engineers, who placed little blame on Caltrans, and the state legislature promptly appropriated funding for Caltrans to pursue a dramatically expanded seismic retrofit program. The slippage narrative that ultimately emerged was not articulated in terms of a problem with Caltrans design practice, but rather in terms of inadequate funding for Caltrans retrofit efforts, which could be remedied by throwing money into a massive, statewide retrofit program.

Slippage, Repair, and the Caltrans Seismic Retrofit Program

Thus began what one engineer characterized as a "golden era" in Caltrans bridge engineering, characterized by a large volume of retrofit design proj-ects, rapidly changing design practice, increased interaction with the research community, a deluge of laboratory testing data, and a more collegial envi-ronment that also increasingly valued those with the most up-to-date engi-neering knowledge. To deal with the large volume of retrofit work, Caltrans contracted with major bridge design firms across the state and for large bridges, like the Coronado Bridge, engaged peer review panels of outside engi-neers and researchers to review plans before construction. The retrofit pro-gram itself was a major systemic engineering effort, requiring coordination and prioritization of retrofit efforts across the entire California road system.

In assessing the Caltrans seismic retrofit program of the 1990s from a repair perspective, it becomes apparent that several connected slippages actually were being articulated and repaired during this period. Caltrans engineers articulated the first slippage around their engineering knowledge and design practice following the 1971 San Fernando earthquake. Although it was not immediately clear what the alternative was, Caltrans engineers recognized their understanding of bridge engineering as incomplete and perhaps obsolete, and they gradually developed an alternative way of doing things that repaired the slippage around their engineering methods and theory. This process enabled them to recognize a second slippage— this one between the obsolete material form of existing bridges designed to old engineering standards and the form those bridges should take to meet Caltrans's new design standards. Seismic retrofit was the solution to this slippage, but because Caltrans engineers did not control the resources needed to retrofit bridges on a large scale, this slippage remained only partially repaired through the 1980s. It was not until the public, and political leaders, were able to articulate their own sense of slippage between how they expected bridges to perform and how they actually performed in the 1989 Loma Prieta earthquake that this second slippage was finally repaired. Partly because Caltrans had already begun to address the first slippage by reforming their design practice, this third, public slippage was articulated in terms of a mismatch in the amount of funding Caltrans needed compared to what had been provided for seismic retrofit. The seismic retrofit program of the 1990s became the ultimate solution to all of these slippages, finally leading to a sense of closure and a newly stabilized sociotechnical network around Caltrans bridges and design practice by the late 1990s.

This series of events reveals how repair efforts can continue, with slippages unsettled, for large portions of the lifetime of an infrastructure system before being resolved, if they are indeed ever fully resolved.[30] It also shows how repair efforts serve to stabilize the dynamic sociotechnical networks surrounding infrastructure systems over time by allowing for periodic renegotiations of the relationships among the social, material, political, and discursive elements of these networks in response to changing conditions.

Bringing Seismic Retrofit Back to Barrio Logan

This statewide drama came home to Barrio Logan in 1995 when Caltrans selected the Coronado Bridge for seismic retrofit. With a tight timetable for retrofitting, Caltrans decided it needed to get ahead of concerns about the murals right away, partly motivated by NEPA and NHPA requirements and partly out of a desire to avoid lawsuits that could greatly slow the retrofit approval process.[31] In addition, several people at Caltrans District 11 in San Diego had developed a personal interest in preserving the murals.[32] Although an outside engineering firm would design the retrofit in cooperation with Caltrans structural engineers in Sacramento, the local impacts were the responsibility of District 11.

A diverse team of local employees took responsibility for issues around the murals and park, including engineers, landscape architects, an archaeologist, and an environmental planner. In a move that reflected the unusual degree of community engagement with the Coronado Bridge, they decided to begin outreach efforts immediately, before design options were settled, meeting with local leaders and speaking at public meetings starting late in 1995.[33] They initially presented a range of possible retrofit strategies for the approach ramps, many of which would have had a dramatic impact on the murals—including placing steel jackets around the columns, adding concrete on the sides of the columns, or removing and replacing the existing columns. Initially, it appears that the Caltrans team thought these options might be acceptable to the community and that mitigation measures like paying artists to repaint or modify the murals might satisfy community concerns. But community activists would have none of it: based on their previous experiences, they had no reason to trust Caltrans. An artists' group circulated a newsletter demanding "no retrofit, not now, not ever!" and a new mural appeared on one of the columns in Chicano Park with the prominent message: "No Retrofitting." In sharp contrast to the past, politicians and newspapers took up the cause of preserving the murals.

At the same time, a historian from Caltrans headquarters in Sacramento began his assessment of the park, bridge, and murals to comply with the NHPA, which required him to assess whether they were eligible for the National Register of Historic Places, which would give them protected status. In addition to historical research, he and the District 11 archaeologist interviewed community members early in 1996. It quickly became clear that the park and the murals would qualify for the National Register.[34] Meanwhile, interactions with the Barrio Logan community showed no signs of progress, with the community engaging in protest marches and preparing to resist any retrofit effort.

Caltrans engineers and planners, as well as the firm designing the retrofit, were now in a tough spot. The approach ramp columns clearly needed to be retrofitted, and all of their typical retrofit strategies would have an impact on the murals. But they also faced an empowered community with a history of effective activism and many political allies, who had no trust in Caltrans's good faith or technical judgment. And they were now trying to retrofit a structure that qualified as a historic landmark. Their solution was to draw on the expanded network of allies Caltrans had built around bridges and seismic retrofit since the 1970s and thus reach out to their frequent research collaborators at UCSD.

Frieder Seible, a characteristically blunt but politically savvy engineering professor originally from Germany, headed the structural engineering laboratory at UCSD. Along with steel jacket retrofit expert Nigel Priestley, he was a member of a Caltrans peer review panel charged with reviewing the design of the Coronado Bridge retrofit. Caltrans staff thought he might have more credibility with Barrio Logan community representatives as a retrofit expert and someone less associated with the Caltrans bureaucracy.[35] Perhaps even more important, Seible had a laboratory full of large-scale replicas of concrete bridge columns that had been partially destroyed in various tests simulating earthquake damage.[36] So in mid-1996, at the request of Caltrans, Seible invited community members to UCSD, where he gave them a detailed primer on mechanisms of earthquake damage, discussed retrofit methods, and took them on a tour of the laboratory. Although Seible mentioned possibly putting, in his words, "tin cans" around the columns, which worried some of the attendees, he made an immediate personal connection with community representatives and told them he would help them in any way he could.[37] This had a major impact on the tenor of discussion: the "no retrofitting" stance died down after this meeting, and Caltrans began to have more positive interactions with the community. (In fact, a damaged test specimen from the UCSD laboratory is now on display at the park.)

Following this meeting, Seible pushed Caltrans to develop a retrofit solution that addressed community concerns. As a peer reviewer, he suggested design engineers consider nonstandard retrofit options. Also, in a stroke of luck, soil tests in Chicano Park turned out better than expected, significantly reducing expected seismic forces on the bridge. With Seible's encouragement, the designers ultimately came up with a strategy of retrofitting only the underground footings of the columns and their top connections to the roadway, almost entirely sparing the murals. Seible presented this strategy at a meeting of the Chicano Park Steering Committee in December 1996 (attended by Sims) to applause and a palpable sense of relief among attendees. In the end, the Coronado Bridge retrofit was completed successfully with no unexpected damage to the murals. Since then, the murals have become increasingly recognized as a major cultural resource in California and nationally. Based on research conducted by Caltrans cultural resource analysts and local activist Josie Talamantez, the park and murals were ultimately added to the National Register of Historic Places in 2013 and were designated a National Historical Landmark in 2016.[38]

This resolution of the Chicano Park murals retrofit controversy illustrates an important point about repair: like any design process, it involves what STS researcher John Law has called "heterogeneous engineering."[39] Technical problem solving alone is rarely sufficient to enact a successful repair. In this case, Caltrans had to carefully negotiate among community interests, politicians, retrofit technologies, peer review panels, soil tests, and an engineering laboratory to weave together a solution that made the retrofit possible from both technical and political perspectives. This work involved not only engineers but also historians, planners, and archaeologists. The previous efforts of Barrio Logan to repair their devastated community also invoked a heterogeneous assemblage of protest marches, activism, political maneuvering, landscaping, historical preservation laws, and artwork. When Caltrans decided to retrofit the bridge, artists and activists in the community were able to use the new order they had woven around the bridge since its construction to force Caltrans to delicately renegotiate its own position in the sociotechnical network to complete the project. Everything from earthquakes to rebar to archaeology to the Chicano civil rights movement was drawn into this process. By engaging in this heterogeneous engineering, participants once again renegotiated the sociotechnical networks around the Coronado Bridge, stabilizing relationships between the bridge, engineering researchers, Caltrans, local politicians, and Barrio Logan in a new configuration.

The association of repair work with care work is also relevant here. Community activists and artists engaged in a type of care work by taking responsibility for preserving the murals—in this case, including the emotional labor of engaging with Caltrans staff in a way that built a sense of shared interest in preserving the murals. Caltrans staff—engineers as well as planners, archaeologists, and others—also put a great deal of time and effort into their interactions with community activists. Much of this effort was focused on attempting to manage community sentiments and demonstrate that Caltrans cared about those sentiments. At one level, this emotional labor served Caltrans's organizational interests in removing obstacles to the retrofit project, meaning that Caltrans personnel were required to demonstrate an attitude of caring whether or not they actually felt that way, much like the flight attendant managing the overheated passenger in chapter 2. However, a number of Caltrans personnel appear to have developed a genuine feeling of connection with the Barrio Logan community and a personal interest in preserving the murals, a crossover of personal and work concerns that is again not unusual in care work.

The engagement between Caltrans and the community was not just a relationship-building exercise, however; it became an engineering problem as well. This is particularly apparent in the involvement of UCSD professor Seible. He not only connected with the community on a personal, affective level, leveraging his credibility as a researcher; he also translated community concerns into engineering terms and used his role as a Caltrans advisor to influence the design process. The result of this engineering-inflected care work was a retrofit approach that accomplished the needed physical repair to the bridge while also building a more positive relationship between Caltrans and Barrio Logan. This skillful interweaving of emotional and technical labor is very similar to the way the repair technicians in chapter 2 handled their local negotiations around the "cold office problem." This suggests that the caring aspects of repair work can come into play at the systemic level and perhaps beyond.

The Ongoing Struggle for Barrio Logan

Chicano Park has been no magic bullet for Barrio Logan in terms of gaining political power or control over its own fate. Although Barrio Logan now has more political clout than it did in the 1970s, the legacy of racism and disregard for the community is very much alive. In the 2010s, planners and community members put enormous effort into developing a Barrio Logan neighborhood zoning plan that would finally separate residential and industrial land uses. However, owners of shipbuilding and other industrial businesses in the area, who felt the plan would be too restrictive, used their political connections to get it put to a city-wide vote in 2014. In an upsetting development for Barrio Logan residents, the rest of the city rejected the plan.[1] In addition, the murals

---

The Ongoing Struggle for Barrio Logan (continued)

have been subjected to vandalism for years, and the park has recently become a target for disruptive protests by right-wing "Patriot" groups that object to the existence of the park and murals.[2] As one of the few relatively inexpensive neighborhoods adjacent to downtown San Diego, Barrio Logan also appears increasingly vulnerable to gentrification. When we visited the neighborhood in January 2019, we noted a number of new coffee shops, breweries, and art galleries, with a new light rail station adjacent to the park providing increased access. Activists have recently protested plans to build an enormous football stadium and convention center nearby.[3] These ongoing issues illustrate some of the complexities and limits of infrastructural repair as a strategy for maintaining power in a constantly evolving political and urban landscape. The "No Retrofitting" mural remains in Chicano Park, looking as if it could have been painted yesterday.

---

1 Joshua Emerson Smith, "San Diego Redrafting Highly Contested Blueprint for Barrio Logan," *San Diego Union-Tribune*, August 28, 2017, https://www.sandiegouniontribune.com/news/environment/sd-me-barrio-logan-update-20170825-story.html; Julie Stalmer, "Barrio Logan Community Plan Not Updated Much," *San Diego Reader*, September 23, 2017, https://www.sandiegoreader.com/news/2017/sep/22/stringers-barrio-logan-community-plan-not-updated/.
2 Sandra Dibble and Kristina Davis, "Officer Punched, Tensions Flare at 'Patriot Picnic' at Chicano Park," *San Diego Union-Tribune*, February 3, 2018, https://www.sandiegouniontribune.com/news/politics/sd-me-patriot-picnic-20180131-story.html.
3 Dave Rice, "Barrio Logan Group Says Basta to Stadium Rhetoric," *San Diego Reader*, July 15, 2016, https://www.sandiegoreader.com/news/2016/jul/15/ticker-community-fighters-calls-bull-t-stadium/.

Conclusion: Infrastructure, Power, and Materiality

In this chapter, we have explored how materiality and power converge in the design and repair of infrastructure systems through a case study focusing on several time periods in the history of the Coronado Bridge and bridge engineering in California. The case study covers three major episodes of repair: the takeover of Chicano Park and painting of murals, Caltrans engineers' efforts to reform their design practice and retrofit bridges, and the negotiation between engineers and Barrio Logan community members to make seismic retrofit of the Coronado Bridge possible. This set of episodes is relevant because it encompasses a diverse range of actors, radically different types of repair, and both relational and systemic scales of infrastructure repair. Following how these diverse elements come together through a period of several decades allows us to see how these networks can evolve over time through repair. It also shows how local networks (like those surrounding the park and murals) and larger-scale networks (such as the political and technical networks surrounding bridge design methods at Caltrans) can drift apart and become reconnected over the lifetime of an infrastructure system, often during episodes of repair.

One thing this history shows is how materiality and discourse interact in repair. In each episode in this chapter, we see a movement from changes in material circumstances to discursive articulations of slippage and repair; in turn, negotiations over slippage are realized in material form through repair. In the first episode, the Barrio Logan community responded to the intrusion of the bridge into their neighborhood by articulating a complex set of grievances and cultural associations, centering around a sense of community investment in the beleaguered barrio, and connecting to larger discourses of civil rights, Chicano identity, and the relationship of culture, political power, and land. This discourse was also symbolically realized in the material form of the park and murals, which illustrated community history and values in visual form. This represents an unusually close convergence between material and discursive forms.

In the second episode of repair, a material disruption—the 1971 San Fernando earthquake—spurred Caltrans engineers to develop a discourse of obsolescence around their design practices, which led them to new engineering knowledge and design methods that connected to an expanded network of allies in the research community. The new knowledge and methods derived from this expanded network were then realized in new bridges. Dealing with existing bridges was a more difficult problem, because the concrete, material form of these bridges was resistant to change. In the end, the steel jacket method made it possible to adopt the required material changes with minimal internal modification to bridges. This made seismic retrofit of the bridge columns possible and resolved the discourse of slippage that had surrounded Caltrans engineering practice. Finally, the retrofit of the Coronado Bridge approach ramps represented a convergence of these two slippages: Caltrans articulated a need for retrofit, and the Barrio Logan community rearticulated the connection between community identity and Chicano Park. These views initially appeared to be in stark opposition to one another. But through a series of subtle political negotiations, exchanges of expertise, and detailed engineering studies, a material configuration for retrofit emerged that could effectively reconcile the discourses of the two communities. This reconciliation is now part of the material form of the bridge.

The episodes of repair in this chapter also illustrate the complex and sometimes ambiguous relationships between infrastructures and power structures, driven by the way large-scale infrastructure systems are inevitably embedded in a variety of localized settings. Infrastructures tend to reflect the centralized power structures of the day, because powerful interests are often represented in their planning and design, while communities marginalized by political exclusion, distance from centers of power, racism, and economic inequality are not. Infrastructure systems themselves often exacerbate this marginalization. Somewhat paradoxically, however, the fact that infrastructure systems often are materially embedded within marginalized communities and spaces can occasionally give communities a point of political leverage. Because the Barrio Logan community originally had little political power, its residents were less visible to planners and engineers and excluded from the sociotechnical networks that came together to get the bridge built. As a result, the configuration of the bridge did not reflect their interests. By gaining control over the portion of the bridge in their neighborhood and materially transforming it into a work of art, they were able to connect their interests to new cultural preservation laws. When it came time to retrofit the bridge, they translated these connections into a place of limited but nevertheless real power within the networks surrounding the bridge, which they used to ensure their concerns were addressed.

This is not just a story of community empowerment, however: by responding to the concerns of the community and more fully incorporating Barrio Logan into the sociotechnical networks surrounding the bridge, Caltrans also benefited by gaining some control over one of the last remaining sources of conflict over the bridge. The result is a bridge that is better adapted to both earthquakes and the evolving political and social environment. This is an example of how repair enables infrastructure systems to overcome potential disruptions by moving to new, stable states, ensuring their persistence over time.

4 From Versailles to Armageddon: Building and Maintaining the Infrastructural State

Introduction: An Infrastructure for Nuclear Armageddon

During the twentieth century, the United States detonated more than one thousand nuclear weapons. Aside of course from the two bombs exploded over Hiroshima and Nagasaki, each of these weapons was exploded to test and refine designs developed over decades of work in the post–World War II era by US weapons designers, engineers, and technicians. Testing nuclear devices not only allowed weapons designers to check the results of their models, but also sent a very public message to other countries about the nation's atomic capabilities. In this way, the mushroom cloud demonstrated the organizational and technical capacity of a vast and secret weapons infrastructure. Behind the hundreds of tests stood a weapons complex of research, administrative, and manufacturing work, comprising sites across the country to design, test, and create the bombs. Los Alamos, New Mexico, where the first atomic bombs were designed and built, is the best known of these locations, but the nuclear weapons complex also included uranium and plutonium enrichment facilities at Oak Ridge, Tennessee, and Hanford, Washington, respectively, and other major research, production, and testing sites spread throughout the country. All of this is to emphasize that the systems for producing nuclear weapons might not be the first example of an infrastructure that jumps to one's mind, but producing a reliable bomb—one that appeared as a credible threat to geopolitical rivals—required a massive system of coordinated material, technical, and organizational resources. In turn, the complex provided employment, prestige, and stability for many thousands of US scientists, engineers, technicians, and administrators.[1]

There is nothing subtle about detonating an atomic bomb. Even underground testing sends a distinct seismic message through the ground that other countries can detect and register. During the Cold War, nations including the United States, the Soviet Union, the United Kingdom, France, and China demonstrated their capabilities and joined the group of global nuclear powers through testing; more recently, India, Pakistan, and North Korea have used testing to signal their arrival in this system of power and the terror of nuclear Armageddon. Testing was considered an essential element of nuclear deterrence, where exploding a bomb was meant to, according to one weapons designer, "scare small children" through the specter of nuclear war.[2]

The Cold War system of nuclear weapons design and production was thrown into uncertainty after 1996, when countries around the world signed the international Comprehensive Test Ban Treaty, forbidding the testing of nuclear weapons. In the United States, as in other nuclear weapons states, this left many thousands of employees unsure of their futures and the prospects for their work in the nuclear weapons complex. It also opened up new questions about how to position nuclear weapons as a credible deterrent in the absence of the routine demonstration of their capabilities through testing. Within the complex, scientists and administrators sought new methods for securing the reliability of the existing stockpile of US nuclear weapons, the expertise of weapons scientists, and the future of the complex itself.

How would weapons scientists, military brass, and politicians continue to convey the threat of a US nuclear arsenal without testing and, at the same time, preserve the far-flung infrastructure of the complex? This question—about the fate of the US nuclear weapons complex and its search for legitimation at the end of the twentieth century—helps us see the scale and the stakes for this chapter, where we explore the role of infrastructural repair at the level of nation-states and the geopolitical and economic interests that construct and maintain national and international structures of power. Repair on this scale involves work on physical structures and systems, the kind of materially based repair required to dig a canal or fix a bridge, but it also takes us into realms where the repair of symbolic and discursive systems goes hand-in-hand with the maintenance of materiality. Who controls and benefits from infrastructures and the economic, political, and symbolic resources that may flow from them? In many ways, these questions and the stories we use to explore them are a direct continuation of our discussion of systemic repair featured in chapter 3; seeing the way that infrastructures span nations and connect the globe will also help lead into our analysis of global infrastructures in chapter 5.

To get from the systemic to the global, however, it is critical to see just how important the building of infrastructures has been for political states of all kinds, from early urban settlements to colonial empires and contemporary nation-states. In this chapter, we focus mainly on the latter two forms of state formation, covering roughly from the period of empire building and colonization by Western powers in the seventeenth century to the era of nations and nationalism that typify the world's political structure today. Infrastructures form a key source of power for states in this period, where material systems for controlling land, populations, and resources reinforce the power of the state and serve as sources of authority and legitimation for a range of persons through the policies and resources supported by infrastructures. Specific interest groups may benefit or be marginalized through the design of infrastructures and how they are repaired, especially via the economic, political, or cultural capital that rulers, bureaucrats, commercial interests, military leaders, and engineers and other experts accrue through infrastructures. Once those investments and benefits are in place, literally built in to the sociotechnical character of infrastructures, repair becomes a matter of personal and class interest for the elites who develop and control infrastructures, meaning that those who benefit from them are more likely to favor an approach to infrastructural repair that maintains those interests.

This *repair as maintenance* may act on the concrete, steel, and other material elements of infrastructures, but also through ideology and identity, the kinds of cultural meanings that we invest in these systems. We pay special attention to the interplay of materiality and discourse here, analyzing cases that show both how expertise is constructed through infrastructures and the struggles that politicians, scientists, engineers, and others face when infrastructures are challenged and disrupted. For example, in the case of the US nuclear weapons complex in the late twentieth and early twenty-first centuries, the material and technical capacity for designing and deploying nuclear weapons was tied to immense economic and political structures, each in turn seen as critical for convincing the rest of the world that the United States had the power to obliterate its geopolitical enemies and rivals.

The loss of testing called several of these interconnected elements into question, thereby initiating a sense of urgency among physicists and engineers, military leaders, and defense hawks; questions about the future of the system threatened not only their jobs but also the legitimacy of the weapons complex. (We return to this case in more detail later in this chapter.)

In sum, to understand how states have developed in the past several centuries of world history, it is important to see how they are *infrastructural states*, that is, political bodies with embedded interests in maintaining structures of material and discursive power.[3] Whereas in chapter 2, we emphasized the role of the human body in sensing the need for repair, here repair is initiated and sponsored by those who gain capital from the structure and durability of the infrastructural state. In this chapter, we use the umbrella term *elite* to refer to the persons and interest groups that benefit most directly from infrastructures and have the strongest incentives to protect them through repair. The administrator overseeing a large nuclear weapons research facility, a politician demonstrating their influence through a showy new building project, and a developer hoping to make a sizable profit from investments in land may not be uniform in terms of their access to and control of the infrastructural state and its levers of power, but each has an outsized role in the creation and maintenance of these systems. As we will see, infrastructures and their repair have had a key role to play in the establishment of our global political order, especially as government, military, and commercial interests developed and maintained infrastructural systems to achieve their goals and consolidate power through sociotechnical means.

While the infrastructural state provides considerable resources and power for an elite who controls and benefits from this system, this does not mean that they wield total and unquestioned control. We also detail some examples where those subject to infrastructural systems resist and rebel against these structures. Infrastructures can be used to divide the haves from the have-nots, supporting an elite at the same time that they disenfranchise those who might have limited access to critical resources like food, water, or energy and the infrastructures that provide those goods. As in the case of the Barrio Logan neighborhood, sometimes disadvantaged communities have an unwelcome infrastructure built right on top of their homes. Those same structures, however, can serve as a means for communities to employ the materiality and symbolic power of infrastructures for their own ends, potentially wresting their own sources of power from sociotechnical systems.

This means that repair at the scale of states and populations is not always about fixing literally broken infrastructures, though of course that is sometimes the case. Instead, we focus in more depth on the ways that infrastructures are built and maintained to embody systems of power. Who has infrastructural power, how did they get it, and how do they keep it? We investigate three key themes related to the infrastructural state: the importance of land and place for initially creating these structures of power, the rise of expertise and systems of expert knowledge through infrastructures, and the communities and identities that grow in tandem with infrastructures and groups of people on a national scale.

In the course of exploring these three themes, we also trace the role of three key sets of human actors negotiating the design, implementation, and repair of the infrastructural state: elites, experts, and the populations subject to the infrastructures built and maintained by the first two groups. While the boundaries among these three sets of actors are blurry, the general interests and activities of each group help us see how repair at the level of states plays out through the structures, organizations, and identities that are built into infrastructures. Modern elites became "elite" in part through the creation of the infrastructural state, deriving power through systems that helped them control land and other resources. Cultivation of experts and new systems of knowledge and technique were and are a key part of these building projects, and experts themselves can participate in this system of power and privilege through the sponsorship of elites. However, experts' grasp on these resources is contingent on their ability to support the maintenance of infrastructures, meaning that they play a key role in repairing the stability of the infrastructural state in times of change and crisis. When things break down, elites count on experts to serve as their repair workers.

The scope of infrastructures and their connections to the state mean that we are all subject to these structures and have a stake in both their positive and negative impacts. The risks and benefits of infrastructure are not evenly distributed, however. While some populations consistently benefit from building and using infrastructures, others are more vulnerable to the risks and negative impacts they create. The rise of the infrastructural state means that countless humans have been abused, conscripted, marginalized, and even subject to injury and death via infrastructures; the exercise of power through infrastructures almost always comes with costs that are not equitably borne by the experts and elites, who accrue benefits from them. The very materiality of infrastructures, however, ensures that they may provide potent targets of protest and reappropriation, meaning that the power of elites is never total or unchanging. The struggle among these three groups as they seek to control and repair infrastructures to support often divergent interests is an important theme of this chapter.

Building the Infrastructural State: Turning Land into Systems of Power

Many of the infrastructures that we have described in this book have one thing in common: they are grounded in a specific place, their material form part of a landscape, a community, or even a region. Infrastructures such as bridges, canals, and roads use geographic topography and features to channel the power of nature and achieve human goals, creating sociotechnical systems that become new landscapes in and of themselves.[4] These kinds of infrastructures appeared early in human societies and had an important role in supporting complex political states. In some cases, such as Egyptian irrigation systems and Roman aqueducts, the close historical connection between infrastructure and the rise and fall of state power is well known, and even taught to children in school. Recent studies suggest that infrastructural harnessing of nature is a key part of the history of a much wider range of civilizations around the globe.[5] So although state investment in infrastructures is nothing new, scholars who study the role of infrastructures in supporting modern states emphasize a set of transitions that have taken place over the past several hundred years. They include the rise of capitalist economies and emerging global markets, the colonial ambitions of the West, scientific and technological advances and the widespread exchange of ideas through printed papers and books, and patterns of inequality and racial ideologies that structured and legitimated who had, and did not have, resources and rights.[6] While the roots of capitalism, colonialism, and emerging scientific and racial ideologies may have earlier foundations in major historical changes such as plagues and shifting systems of religious thought,[7] an era of acceleration starting in the seventeenth and eighteenth centuries intensified the growth of the modern infrastructural state, increasingly tying people and materiality together via sociotechnical systems. These were early steps toward the infrastructural engineering of the earth itself, a topic we discuss in more detail in chapter 5.

The control of territory through infrastructural systems allowed states and state actors to partner with the emerging interests of capital and colonialism, providing a synergy of interests between political and economic elites and spurring the creation of new systems of expert knowledge and practice. In this way, reengineering land through infrastructures created not only the material resources to support the accumulation of capital and power but also a context in which new or sharpened tools, such as censuses, mass media, and collective imaginaries, allowed the state to study and manipulate its human populations, develop new centers of taxation or profit, and generate a sense of allegiance or at least acquiescence to a centralized infrastructural state. While not a complete explanation for how modernity became modernity, the control of land and resources based in earth and water helps us see how the power and resources found in natural systems were transformed into built environments that facilitated the accumulation of social and technical power. In turn, those who benefited from these structures developed new interests in the resources that flowed from them, helping us to see how an approach to repair as maintenance is often favored by elites and experts. In the face of challenges, crises, and even revolutions, repair as maintenance is about protecting power, and this section shows how this approach to repair was effectively built into the landscapes and institutions of the early modern European state.

One way that infrastructures facilitated this set of transformations was via their capacity to control time and space. Space-time compression, a concept developed by geographer David Harvey, highlights the roles of canals, roads, railways, bridges, and ports for opening up new avenues for commerce, military conquest, and the imperial ambitions of states, compressing the distance between any two points on the earth and making time itself subject to sociotechnical control.[8] Infrastructures that reduced the time to move troops, bring goods to market, and share information effectively shrunk the globe, and those who created and controlled these infrastructures reaped the benefits of mastering space and time. But a focus exclusively on how infrastructures enable the movement of people and things misses the important and material ways that the administration of land via infrastructures increasingly became the goal of states in the early modern period. Controlling territory was a key means of developing new capital through the emerging infrastructures that helped early modern states take control of and harness these resources.

Sociologist Michael Mann, whose multivolume analysis of the sources of social power provides a key resource for understanding the infrastructure-state relationship, emphasizes that infrastructural power is a key goal for political states because they are bounded by a geographic limit, and infrastructures help them achieve "a unified territorial reach."[9] By combining access to the material boundaries of a place with the administrative capacities of modern states (bureaucratic functions such as surveying, taxing, and mapping), Mann argues that political elites wield considerable power:

> [The state's infrastructural] powers are now immense. The state can assess and tax our income and wealth at source, without our consent or that of our neighbours or kin (which states before about 1850 were never able to do); it stores and can recall immediately a massive amount of information about all of us; it can enforce its will within the day almost anywhere in its domains; its influence on the overall economy is enormous; it even directly provides the subsistence of most of us (in state employment, in pensions, in family allowances, etc.). The state penetrates everyday life more than did any historical state. Its infrastructural power has increased enormously.[10]

By compressing time and space and extending their administrative and technical reach, these new capacities of the state allowed rulers to develop the "infrastructural power" that Mann describes, and in the remainder of this section, we treat two scholars' case studies of the early modern infrastructural state. Sociologists Chandra Mukerji and Patrick Carroll demonstrate through historical analysis of the growing infrastructural state in seventeenth-century France and Ireland, respectively, how a growing administrative state, technology, ideology, and competition for land and other resources all came together to fuel what Mukerji terms the "territorial ambitions" of early modern European states. In one of her key works, Mukerji focuses on the creation of the French gardens at Versailles under the direction of Louis XIV and his chief advisor, Jean-Baptiste Colbert, in the latter decades of the seventeenth century; a second book details the construction of France's Canal du Midi in the same period, a 240-kilometer waterway that helped connect France's Atlantic and Mediterranean coastlines, and was considered a marvel of engineering at its construction.[11] While a large-scale canal building project and design of a royal garden may not seem closely connected at first glance, Mukerji emphasizes that each project supported Louis XIV's attempts to consolidate power centrally through the control of territory both in France (where the state sought to bring provinces under greater centralized control) and out of France (through colonization). State infrastructures demonstrated the power of the monarchy and brought increased bureaucratic and military oversight to sites far from the seat of power, ultimately centralizing state administration and creating the symbolic and material apparatus of the infrastructural state.

Symbolic and material power were embodied in the plan for Versailles, which Mukerji emphasizes was created as a *place*, deliberately engineered to embed assumptions about French political and military power in the land itself. This focus on land and place is important to highlight. Infrastructures such as roads, canals, harbors, and military installations facilitated the movement of people, commercial cargo, and materials of war, but they also remade landscapes and ecologies according to these interests.[12] As Louis XIV sought to compete with European rivals as well as unruly nobles in his own court, Versailles provided the setting to "[speak] obliquely but effectively to French prowess in war."[13] Innovations in irrigation, hydraulics, and other forms of engineering were needed to support the display and spectacle of the gardens at Versailles; in this way, Versailles had a symbolic power founded in the resources marshaled to construct and maintain such a vast garden, but it also generated new knowledge and technologies that hybridized with military engineering such as design of forts and battlements. The geometric layouts and designs of the gardens echoed and influenced military installations, meaning that Versailles served both symbolic and practical uses in the emerging era of regional and global ambitions for empire (see figure 4.1).[14]

Whereas Versailles and its regal collection of opulent chambers, gardens, and fountains signified the centralized power of the infrastructural state, the construction of the Canal du Midi illuminates how the mastery of hydraulic engineering allowed Louis XIV to extend control of the French state into territory far from his palace. Although some of the same engineering knowledge that supported the creation of Versailles was used for the Canal du Midi, Mukerji's history of the project shows how its completion was achieved only through the appropriation of knowledge developed by the rural peasants who had long lived along the rivers and streams

that became the path for the canal. While the canal was deemed a feat of "impossible engineering" due to the demands of channeling water through the Pyrenees, local residents had developed practices and technologies for controlling the flow of water through the mountains and employing it for their own purposes via mills, laundries, and weirs for the retention of water. Mukerji argues that by latching onto the ideas of these "indigenous engineers," many of them women, the canal builders were able to solve many challenges of routing the canal through the mountains while also rendering their contributions invisible.[15] Because the peasants were largely illiterate, their work and knowledge were not recorded in the history of the canal but instead attributed to the genius of the canal's chief builder, Jean-Mathias Riquet, and the glory of Louis XIV.

While Mukerji emphasizes the ways that the canal's construction and use was flawed—sections quickly failed under the challenges of the massive built system—the way that these failures and the contributions of local knowledge were papered over helped to create an aura of impersonal and inevitable governance through the infrastructural power of the state. Like many of our encounters with infrastructures, the canal was far from perfect in its execution and operation, but an emerging modern French state was built in part on the perception that it could administer projects on the scale of the Canal du Midi with rational and powerful techniques of control.[16] Materially, the canal was constructed with the support of indigenous engineering knowledge, which was in turn used to build the symbolic mastery of expert systems and the central state over the harshest landscapes in its domain. The challenges that the state's engineers and administrators faced

Figure 4.1
Images of the Latona and Apollo fountains at Versailles. Mukerji notes: "[The Latona and Apollo fountains] spoke eloquently to the military might and ambitions of the king and state. The central statue of the Latona fountain showed a delicate and beautiful young mother with her children (one of which was Apollo as a baby) surrounded and attacked by quite horrifying figures of angry peasants, frogs, and lizards. . . . It was an image of vulnerability that was balanced in the Apollo fountain by an image of complete and utter adequacy. . . . The counterpoint of the two fountains, contrasting youthful vulnerability and adult potency, provided an interesting comment on the Sun King's [Louis XIV's] coming of age." Quotation from Chandra Mukerji, *Territorial Ambitions and the Gardens of Versailles* (New York: Cambridge University Press, 1997), 68–70. Images by Christopher R. Henke, 2019.

were repaired using local knowledge, providing a foundation for both engineering science and the state's legitimacy and power.

Extending the infrastructural state deep into a landscape facilitated the appropriation of knowledge, as in the case of the Canal du Midi, and at the same time allowed state actors to survey and catalog the material and human capital contained within specific regions. Patrick Carroll describes a similar exercise of state infrastructural power in the case of seventeenth-century Ireland, where William Petty represented the English state and its attempts to domesticate Irish lands and peoples during Oliver Cromwell's invasion in 1649.[17] Carroll details the use of maps, surveys, censuses, and other means of collecting data to understand and ultimately control Ireland as a colonial territory of the growing English empire. The triangulation of these various forms of data created the "political arithmetic," in Petty's terms, to implement systems of taxation and other sources of revenue.[18] The early foundations of modern science's experimental method were being codified by Petty, Robert Boyle, and other members of the English Royal Society during this period, but these methods were not isolated in laboratories or the salons of the elite. Instead, Carroll shows how the scientific revolution was entirely tied up with political and economic revolutions and the colonial ambitions of European powers. Petty himself was granted an estate of more than 200,000 acres of Irish land as a reward for his efforts in securing Ireland for the Crown.[19]

Coercing land and people into subjects of the modern state helped consolidate considerable social and material power for rulers and their advisors like Louis XIV, Jean-Baptiste Colbert, Oliver Cromwell, and William Petty. While their power ebbed and flowed with the political revolutions that rocked Europe and the rest of the world during this period, the process of transforming land and building the infrastructures of the modern state set in place a network of sociotechnical structures for the accumulation of economic, political, and cultural capital—and an incentive to keep this capital flowing. State-of-the-art infrastructures such as canals helped promote commerce through the sixteenth to eighteenth centuries, setting the basis for the capitalist, military, and citizenship models that undergird modern states. Just as rulers and ruling parties came and went, these infrastructures were often supplanted by newer systems (railroads superseding canals, highways diminishing the importance of trains), but each enabled power from diverse sources and defined the interests of a political class that wanted to maintain their access to power and capital through these structures.[20]

The techniques and tools needed to build these infrastructures were innovative and cutting-edge technologies at the time, highlighting the ambition of those who took on projects even in the face of what seemed to be impossible engineering challenges. High risks led to great rewards once these infrastructures were built into the geographic and administrative systems of the emerging modern state; in turn, those structures needed to be maintained and protected once the means were in place to reap the benefits of access to new resources. In effect, elites and their experts were captured by their own investments of power in and from infrastructures, tied to the structures that supported their positions and sources of capital. Elites' interests in repair therefore were and are built in to both the material design of infrastructures and the administration of infrastructures through state agencies, policies, and practices. Both the modern state and the class of economic and political elites who control and benefit from the state have vested interests in these structures, helping us see the impetus for repair as maintenance at this larger scale.

Embedded Interests: Levees and Flood Control in New Orleans = Repair as Maintenance[1]

Historian Peirce Lewis describes the city of New Orleans as "impossible but inevitable," emphasizing the promise of its location at the mouth of the Mississippi River while recognizing the challenges in siting a city in the middle of an enormous floodplain.[2] Given its geographic potential for trade and military defense, New Orleans's location was just too good to pass up, but how to keep it dry? Beginning in the nineteenth century and continuing largely until today, New Orleans has been protected by an extensive levee system that acts as a "jacket" for the Mississippi and the other bodies of water that surround the city. As more and more of the Mississippi was jacketed in levees with ever higher walls, the constrained river actually created higher waters during flood stages and increased the risk of catastrophic flooding due to breached levees. Nevertheless, the levees-only policy was favored by elites, especially landowners and developers who sought to expand settled areas onto land that was once considered swamp.[3]

Once the land was developed, it became difficult to turn back the clock, and the extensive levee system formed a built environment that reflected the economic interests and power structure of New Orleans, creating a sociotechnical structure for future action. This process was illustrated in the extensive flooding along the Mississippi in 1927, when New Orleans was threatened by rising floodwaters along the city's levees. The commissioners of the regional levee boards and top brass in the US Army Corps of Engineers wielded incredible authority in the face of the flood, conscripting an army of disenfranchised

Embedded Interests: Levees and Flood Control in New Orleans = Repair as Maintenance (continued)

labor (many of whom were African Americans working on post-Reconstruction plantations) to shore up the levees.[4] As the floodwaters continued to rise, city elites conspired with the state to dynamite levees south of the city, in St. Bernard Parish, to serve as an outlet for the rising waters and protect the city from flooding. Nearly ten thousand residents of St. Bernard Parish were evacuated, and their homes and livelihoods were destroyed as the parish became an enormous holding reservoir.[5] As an extreme form of repair as maintenance, this strategy kept New Orleans dry, preserving the structure of both the levee system and power relations in the city, but at the cost of destroying St. Bernard Parish.

Fast-forward to the devastating impact of Hurricane Katrina on New Orleans in 2005, and the repair-as-maintenance approach helps us see the perverse way in which the destruction that the hurricane caused was essentially engineered into the built environment of the city and its river.[6] Death and damage due to a strong storm like Hurricane Katrina was widely predicted many years before the events of 2005, yet the inadequate levee system was not upgraded or replaced with a superior system. The impacts of Katrina were disproportionately distributed, and low-income communities of color faced death and displacement in numbers that reflected the inequities of urban spaces in the United States, as well as the more specific infrastructural ecology of New Orleans and its levee system.[7]

1 The case study in this section is derived from Christopher R. Henke, "Situation Normal? Repairing a Risky Ecology," *Social Studies of Science* 37, no. 1 (2007): 135–142.

2 Peirce F. Lewis, *New Orleans: The Making of an Urban Landscape*, 2nd ed. (Santa Fe, NM: Center for American Places, 2003), 19.

3 Ari Kelman, *A River and Its City: The Nature of Landscape in New Orleans* (Berkeley: University of California Press, 2003).

4 John M. Barry, *Rising Tide: The Great Mississippi Flood of 1927 and How It Changed America* (New York: Simon & Schuster, 1998), chap. 14.

5 Gay M. Gomez, "Perspective, Power, and Priorities: New Orleans and the Mississippi River Flood of 1927," in *Transforming New Orleans and Its Environs: Centuries of Change*, ed. Craig E. Colten (Pittsburgh: University of Pittsburgh Press, 2000), 109–120; Kelman, *A River and Its City*, chap. 5.

6 William R. Freudenburg, Robert B. Gramling, Shirley Laska, and Kai Erikson, *Catastrophe in the Making: The Engineering of Katrina and the Disasters of Tomorrow* (Washington, DC: Island Press, 2009).

7 Joan Brunkard, Gonza Namulanda, and Raoult Ratard, "Hurricane Katrina Deaths, Louisiana, 2005," *Disaster Medicine and Public Health Preparedness* 2, no. 4 (2008): 215–223; Elizabeth Fussell, Narayan Sastry, and Mark VanLandingham, "Race, Socioeconomic Status, and Return Migration to New Orleans after Hurricane Katrina," *Population and Environment* 31, no. 1–3 (2010): 20–42; Carl Bialik, "We Still Don't Know How Many People Died Because of Katrina," *FiveThirtyEight* (blog), August 26, 2015, https://fivethirtyeight.com/features/we-still-dont-know-how-many-people-died-because-of-katrina/.

Experts and Governmentality: The Maintainers of State Power

If repair as maintenance is the preferred approach for those who gain capital from established infrastructural systems, that fact begs a question: Who does all this maintenance work? Historians Andy Russell and Lee Vinsel ask us to recognize and appreciate these often behind-the-scenes workers they term "the maintainers"—those who perform "undervalued forms of technological labour"—who support our everyday sociotechnical orders.[21] While we have explored this work in previous chapters, in this section we focus specifically on those maintainers—experts and their systems of knowledge—sponsored by the state to create and maintain infrastructural systems. Experts and infrastructures have a symbiotic relationship in that it is hard to imagine the existence of complex infrastructures without the technical knowledge and skill required to build and repair them. As we saw with the Canal du Midi, ambitious infrastructural projects often require or inspire the creation (or appropriation) of new knowledge systems and even whole new professions, and so the relationship between infrastructures and experts reflects the splintered division of labor in industrialized societies, where jobs are ever more specialized.[22]

In addition, infrastructures are of much interest to the state, so the symbiosis of infrastructures and experts does not occur in a political vacuum. Instead, state interests shape and support the growth of knowledge and expertise connected with infrastructures. In fact, experts are often part of the state apparatus itself, employed in bureaus of land management, engineering, transportation, agriculture, forestry, and many other specialized agencies. The military ambitions of the state also figure critically here, as research and development funds for technological innovations support many experts with duties related to defense structures and systems. However, this sponsorship comes with strings attached, and experts may feel conflicted about their roles in maintaining the infrastructural state.[23] Communities of experts may also be divided by political allegiances and institutional interests, leading to potential tensions within what we might otherwise consider homogeneous groups of engineers, scientists, technicians, and agency administrators. Overall, experts are deeply embedded within the tangled nest of interests and resources assembled through infrastructures, meaning that they have strong (if sometimes conflicting) incentives to maintain and repair those systems. In turn, state and elite interests support this maintenance and repair because it enables them to continue to benefit from the economic and political capital that accrues through expert control of infrastructures.

Historian and philosopher Michel Foucault's work on the rise of the modern state provides another important set of resources for understanding how infrastructures and systems of expert knowledge coevolved over the past three hundred years. Foucault's perhaps most influential work, *Discipline and Punish: The Birth of the Prison*, is an important touchstone for seeing how these systems not only became embedded within modern institutions of all kinds, but also shifted how we think about human organization and life.[24] As states established the administrative reach described in the prior section, they also increasingly developed tools for shaping the hearts and minds of state subjects. While prisons and systems of punishment might not initially seem important for understanding the connections between infrastructures and experts, Foucault emphasizes the impact of these systems on our methods and discourses of punishment in the modern era and on our everyday thoughts and behaviors.

Foucault begins *Discipline and Punish* with a provocative excerpt describing the execution of a man in 1757 who was sentenced to be drawn and quartered (pulled apart by four horses tied to his various limbs; look it up if you want the literally gory details, but suffice it to say that it is actually a lot of gruesome work to kill someone in this way). Foucault contrasts this method of punishment with an example from just eighty years later, in the early nineteenth century, where incarcerated prisoners are "disciplined" through a highly regimented series of activities, including moral education, labor, and even recreation, all carefully planned to reform criminals and transform them into new people. By emphasizing this contrast between a brutally public punishment *on* the body and a carefully planned discipline *of* the body by penal experts, Foucault highlights the transformation of how states consider and treat their subjects in modernity.

Infrastructures play an important part in Foucault's story, including especially the "panoptic," or all-seeing, structure of modern prisons and other infrastructures of surveillance. Prisons are designed so that guards and wardens can monitor their charges at all times. Continual surveillance, Foucault argues, was and is intended to get into our heads, making not just prisoners but all members of modern states internalize a set of values and practices that lead them to self-discipline. By continually monitoring our own motivations and desires, we can adopt a form of self-control that Foucault argues is, ironically, a more total (and perhaps more tragic) form of discipline than the public spectacle of punishment through torture more common before 1800. In effect, we maintain our own actions through this process of internalizing and acting on assumptions about what it means to behave as a modern member of civilized society. While prisons and other material structures are still important mechanisms where the state can control the bodies of its subjects, repairing the mind, Foucault argues, is ultimately the most effective and invasive approach.

The growth of knowledge systems, especially in human sciences such as psychology and economics, is also central to this shift, as experts developed new techniques and technologies to administer the disciplinary state and were increasingly employed as the bureaucratic maintenance force that did its routine work.[25] *Governmentality* is the term Foucault uses to describe the set of practices and structures that typify the modern state. He defines the state not so much in terms of a power structure—who rules whom—but rather through the mechanisms by which governing is actually done on people's bodies and in their minds. Governmentality includes actual material structures such as panoptic viewing stations and record-keeping systems, as well as the discourses that legitimate these practices, leading to a more subtle but all-encompassing form of power than the direct control of earlier ages.[26]

Foucault helps connect the examples of canals and censuses we described in the previous section, where infrastructures helped states extend their material reach into new territories, with subsequent use of infrastructures to mold citizens and their sense of identity and obligations to the state. For example, Stalin's equivalent of the Canal du Midi, the White Sea–Baltic Canal, connecting the Siberian coast on the northern reaches of the Soviet Union to the Baltic states in the southeast, employed forced prison labor to build the 227-kilometer canal in just two years, between 1931 and 1933. While construction of the canal itself was a key goal of the project, the effort was also meant to support the Stalinist principle of *perekovka*, or reforging, to "mold criminal prisoners into dedicated believers and practitioners of Soviet ideology."[27] In this way, the project was meant to build a sense of nationalist identity and obligation. Among the many experts employed to build the canal, a Writers Brigade of 120 journalists and other authors toured the canal, interviewed the prison laborers, and wrote accounts of the canal "advocat[ing] and applaud[ing] the remaking of Russian society into a Soviet Stalinist society."[28]

Many thousands of workers died in the process of completing the canal, pointing to the limits of the state and its infrastructures to shape the minds of citizens without also policing their bodies, sometimes in brutal ways. Foucault also emphasizes that the subjects of state and expert control do not obey these systems unthinkingly. In fact, he argues, resistance is an embedded feature of power: wherever power is exercised, resistance in some form will be there too.

Transantiago: A Case Study on the Limits of Expert Power

Sebastián Ureta's study of the Transantiago public transit system in Santiago, Chile, explores the impact on commuters when a new and intentionally transformative transit model was introduced to the city in 2007.[1] Ureta uses Foucault's work as a lens to understand how everyday subway riders were treated as *scripts*, or idealized constructions of malleable and polite transit users. While Transantiago was meant to be among the most efficiently planned and executed public transit models of the new century, taking advantage of the newest technologies and planning techniques, the reality of a flood of new users, overcrowded trains, and major delays forced Transantiago's planners to implement a number of policies and practices meant to reorient and "normalize" the commuters' ways of engaging with the system. Signs posted throughout the subway tunnels exhorted riders to be considerate toward each other, security guards were posted on each platform to police behavior, and at peak times of commuter traffic, riders were forced to wait in place before moving to or from train platforms until the system could catch up with demand.[2] While these efforts at repairing users' relationship to and engagement with Transantiago helped salvage the system, Ureta also points to the limits of these disciplinary practices and the larger limitations on experts' and policymakers' attempts to repair the relationship between commuters and the idealized vision of Transantiago.[3]

Ureta's case study helps us see how Foucault's ideas can illuminate the power of discourses when implemented in specific infrastructural contexts, but also the limits of these same techniques. Indeed, just as Foucault emphasizes the subtle but deep way that expert discourses shape our thoughts and actions, he also emphasizes the limits to these systems of power, especially through resistance.[4] When guards must be placed on each platform to enforce

> ideals of discipline, the limited power of internalized discourses is high-
> lighted, demonstrating the ongoing tensions and struggles over infrastructural
> repair.

1 Sebastián Ureta, "Waiting for the Barbarians: Disciplinary Devices on Metro de Santiago," *Organization* 20, no. 4 (2013): 596–614; Sebastián Ureta, "Normalizing Transantiago: On the Challenges (and Limits) of Repairing Infrastructures," *Social Studies of Science* 44, no. 3 (2014): 368–392; Sebastián Ureta, *Assembling Policy: Transantiago, Human Devices, and the Dream of a World-Class Society* (Cambridge, MA: MIT Press, 2015).
2 Ureta, "Waiting for the Barbarians," 604–610.
3 Ureta, "Normalizing Transantiago."
4 Brent L. Pickett, "Foucault and the Politics of Resistance," *Polity* 28, no. 4 (1996): 445–466.

A Fragile Power: The Codependency of Experts and the State

While experts can benefit greatly from their place within state structures, gaining the sponsorship needed to develop new knowledge and create incredible works of material, technical, and organizational sophistication, this patronage is not without a set of trade-offs, given the intense interest that politicians, generals, and investors have in the work of experts and their desires for specific outcomes and payoffs. Chandra Mukerji describes experts' situation as "a fragile power," a term that emphasizes the funda-mental dependency of experts but also acknowledges their place in a system of considerable resources and influence.[29] Mukerji claims that scientists act as a reserve labor force for the state, ready with support in times when the state may be especially in need of their expertise. The state does not neces-sarily have a specific use for every piece of research produced by the experts they sponsor, but instead they see the value in being able to tap a reserve of knowledge and technical skill "more consistently relevant to state interests and visible to government agencies."[30] Mukerji's focus on oceanographic science makes the relevance of these scientists' experience and knowledge for military interests especially clear. The US Navy, in particular, funds a lot of ocean scientists and has a strong yet somewhat detached interest in maintaining a civilian corps of researchers who have significant experience living and working at sea. These skills and systems come into play when the Navy needs to find a lost submarine or when a conflict arises.[31]

Similarly, the US Department of Energy funds a great deal of basic research that is not directly tied to energy or nuclear weapons, but helps maintain its ties to a larger research community and allows the state to project an image of scientific competency in line with its status as a military superpower.[32] As noted earlier in this chapter, the US nuclear weapons complex is an enormous set of infrastructures built around the expert knowledge systems needed to develop and maintain the weapons stockpile. With the end of nuclear testing in the 1990s, scientists and administrators sought new methods of both maintaining and developing the scientific basis for weapons design and securing the credibility of the existing US nuclear arsenal. Ultimately, two key strategies emerged to cope with the loss of testing. The Stockpile Stewardship Program, which remains the basis of US nuclear weapons research efforts, sought to preserve existing knowledge about weapons design primarily through development of computer simulations meant to virtually test the weapons stockpile. A later program that was ultimately canceled, the Reliable Replacement Warhead (RRW) instead focused on the creation of a warhead design that would not require testing to develop and would in turn have a very long shelf life as part of the US arsenal.

These programs were meant to maintain the material destructive capacity of the nuclear weapons arsenal, given a context where the complex's stewards were not able to verify and demonstrate the viability of the weapons through testing. But Stockpile Stewardship did little to protect the interests of the workers, technicians, and machinists who actually built nuclear weapons, most of whom lost their jobs with the end of the Cold War. Instead, it focused on repairing the credibility and status of the technical experts at the weapons research laboratories. Laboratory leaders, in particular, argued that the deterrent effect of nuclear weapons was fundamentally based on the credibility of the experts who designed and created the weapons. Sig Hecker, who directed Los Alamos National Laboratory from 1986 to 1997 and was a key figure in the transition of the weapons complex during the shift away from testing, told us in an interview: "It is the labs per se that provide the deterrence, not the bombs. Bombs can be overcome with newer designs or countermeasures, [but] the Russians will never be able to overcome our ability to evolve and develop new technological capabilities."[33] In Hecker's view, the intellectual expertise and technical resources that stood behind the weapons complex were the true deterrent. Though he is clearly making an argument that supports the interests of the labs and the weapons scientists in the face of a new set of restrictions around testing, he is also thinking through the ways that the work of the weapons complex might be maintained through new missions and a renewed sense of purpose.

In each case, the proposed solutions were meant to repair the credibility of the weapons, as well as the system of scientific expertise behind them. However, the Stockpile Stewardship Program, by arguing for a departure from the design of new weapons and major investment in new computing infrastructures, was resisted by some in the core weapons design community because it was seen as a relatively radical repair of the longstanding discourses and institutions supporting the weapons development process. Although the RRW program might seem on face value to be more transformative, since it championed the development of a new weapons design, discursively and institutionally the program mobilized the kinds of knowledge and skills central to the Cold War design and testing process. This approach appealed to many in the core weapons design community since it seemed to preserve their autonomy and authority within a familiar weapons development regime. These two different ways of framing the technical, cultural, and institutional changes needed to adapt to a radically changed set of political and practical circumstances for work in the weapons complex demonstrate how "repair may serve to reveal hidden differences" among the key stakeholders tied to a complex set of infrastructures.[34] Sponsorship from state and corporate patrons makes experts part of the elite class in their own way, though their status is contingent on shifting demands for knowledge, technology, and repair based on political and economic change.

Infrastructures on Display: Making Nations and Nationalism through Sociotechnical Repair

In the twentieth century, the nation-state became the predominant form of political organization and discourse through the world, and infrastructures continue to be a key part of national strategies for defining legitimacy and power. We have already discussed several examples where sociotechnical systems were and are used to demonstrate the power of the state and develop techniques to materially control and exploit its territories. The era of the

The "Mutual Orientation" of the Internet's Open Culture and Structure

The advent of the nuclear age and the onset of the Cold War in the late 1940s and 1950s created a strategic interest in developing computers and information networks that would allow the US military to monitor nuclear and other activities around the globe. Ideally, these sociotechnical systems would be robust and redundant, able to withstand disruption from attack and sabotage. Thus, many scholars trace the roots of our contemporary internet back to the ramp-up of R&D spending and intensive sociotechnical network development that began during World War II and continued through the Cold War. The open and decentralized—even chaotic—structure of the internet may seem at odds with military interests, ideologies, and organizational approaches. But internet historians such as Paul Edwards argue that the history is more complex and that the structure of systems like the internet actually grew out of the interdisciplinary and relatively nonhierarchical structures of Cold War research organizations like the Manhattan Project and MIT's Radiation Lab.[1] While funding from the state was meant to develop technical capacities, the details and expectations for this work were relatively unstructured, governed by a process of "mutual orientation" where military sponsors, research scientists, and corporate engineers and entrepreneurs shared a common set of political and technological discourses about the aims of US defense research and development.[2] So computers and networking gained their contemporary structure through the organization of knowledge production and technical innovation in the postwar years, not despite it—and now the two of us have the capacity to coauthor this book via Google Docs, watching each other's edits in real time.[3]

Despite the "fragile power" of experts described above, military largesse provided enormous resources for the growth of academic and corporate science during the twentieth century, effectively creating a new class of sociotechnical workers who largely enjoyed lifestyles commensurate with or better than the emerging American middle class.[4] With the advent of internet-based systems of managing and distributing work, however, scholars are increasingly asking questions about the future of this system of expertise. Computer scientist and communication researcher Lilly Irani studies Amazon.com's Mechanical Turk system for distributing "microtasks" to contingent workers around the world—answering surveys, checking the accuracy of web data, and other short tasks compensated by the job, effectively creating an online "market for largely invisible cognitive pieceworkers."[5] In some cases these microtasks might be called a form of repair, where workers are asked to police the content of web advertisements and social media posts.[6] As work is increasingly distributed across the globe, split into ever finer tasks, and compensated with fractional incomes, the human and artificial intelligences that complete this work may increasingly erode the fragile power of experts and other skilled repair workers. The distribution and diffusion of expertise may also make it harder for states to control these knowledge systems and use them for their own geopolitical ends.

1 Paul N. Edwards, *The Closed World: Computers and the Politics of Discourse in Cold War America* (Cambridge, MA: MIT Press, 1996); Paul N. Edwards, *A Vast Machine: Computer Models, Climate Data, and the Politics of Global Warming* (Cambridge, MA: MIT Press, 2010); Janet Abbate, *Inventing the Internet* (Cambridge, MA: MIT Press, 1999); Fred Turner, *From Counterculture to Cyberculture: Stewart Brand, the Whole Earth Network, and the Rise of Digital Utopianism* (Chicago: University of Chicago Press, 2006); Andrew L. Russell, *Open Standards and the Digital Age: History, Ideology, and Networks* (Cambridge: Cambridge University Press, 2014).

2 Edwards, *The Closed World*; Paul N. Edwards, "Infrastructure and Modernity: Force, Time, and Social Organization in the History of Sociotechnical Systems," in *Modernity and Technology*, ed. Thomas J. Misa, Philip Brey, and Andrew Feenberg (Cambridge, MA: MIT Press, 2003), 185–225.

3 Russell, *Open Standards and the Digital Age*, provides a somewhat contrary view, detailing competing standards for networking and the relative failure of democratic methods to adopt a common protocol.

4 Chandra Mukerji, *A Fragile Power: Scientists and the State* (Princeton, NJ: Princeton University Press, 1989).

5 Lilly Irani, "Microworking the Crowd," *Limn*, February 13, 2012, https://limn.it/articles/microworking-the-crowd/; Lilly Irani, "The Cultural Work of Microwork," *New Media and Society* 17, no. 5 (2015): 720–739.

6 Irani, "Microworking the Crowd"; Andrew Russell and Lee Vinsel, "Hail the Maintainers," *Aeon*, 2016, https://aeon.co/essays/innovation-is-overvalued-maintenance-often-matters-more.

nation-state continues this trend, where infrastructures are often deliberately made visible in both a grand way to display the power of the state and a more personal way in which feelings and understandings of citizens' place in a political system are expressed through their relationships with infrastructures. The scope of infrastructures and their potential to transform landscapes and lives provides a powerful means to demonstrate the reach of the state and even define the contours of how people think of themselves in terms of national identities and their place within modern technological and cultural lifestyles. Anthropologist Brian Larkin notes, "Roads and railways are not just technical objects then but also operate on the level of fantasy and desire. They encode the dreams of individuals and societies and are the vehicles whereby those fantasies are transmitted and made emotionally real."[35] Larkin points us toward the ways infrastructures are not just embedded in our landscapes and material environments; they also shape the way we think about ourselves and our place in political structures. In this section, we explore this symbolic power of infrastructures for the state and its citizens in more depth, focusing on two key aspects of how infrastructures shape discourses about social and material structures: through the identities that citizens of modern states develop around their national allegiances and through the efforts of states to make grand statements about their power through infrastructural construction and control. Each of these mechanisms shapes and influences life in modern infrastructural systems and demands various forms of repair from politicians, experts, and everyday citizens.

The first mechanism, nationalist identity, begs the question: What exactly causes anyone in a group of millions to feel like they are somehow the same kind of people as others in their nation? This is the question anthropologist Benedict Anderson tackles in his influential work, *Imagined Communities: Reflections on the Origin and Spread of Nationalism.* As the title implies, Anderson argues that nationalism, or the sense of belonging to a specific nation-state, is an affinity that citizens internalize and *imagine* as a strong and even sacred connection to their fellow nationals. Just because it is an imagined community, however, does not mean it is not a strong one. Indeed, as Anderson powerfully notes, many millions of people gave up their lives during conflicts throughout the past two hundred years in the service of nation-states.[36] How does the nation exert such a powerful pull on our sense of identity and allegiance?

Anderson traces the formation of imagined national communities to some of the same structures and practices we described earlier in this chapter, factors that were also critical to the growth of a modern infrastructural state in contexts such as the English colonization of Ireland. Censuses helped to enumerate populations and make them subject to state control and surveillance.[37] The creation of common languages through the rise of printing and shared media such as books and newspapers standardized previously heterogeneous dialects into a national common tongue.[38] National borders, especially in colonial states, were shaped through the administrative apparatus that the colonizers forced onto the outlines of their territorial aims.[39]

Infrastructures play a role in not only creating the data, media, and organizations that structure and support the modern nation-state, but also, as Larkin notes, shaping the hopes and dreams of citizens. To this we add that infrastructures can be part of our nightmares too, as in the case of the nuclear weapons complex and its capacity for global destruction. Anthropologist Joseph Masco argues that nuclear weapons were the signature technology through which Americans were willing to think through the ultimate national sacrifice during the twentieth century: "The nuclear complex remains a particularly potent national project, informing one way in which citizens imagine both their collective lives and deaths. The unthinkability of the nuclear age has from this vantage point been perhaps *the* American nation-building project since World War II."[40] The United States tested hundreds of bombs, built thousands more, spent trillions of dollars, and sold the narrative that we needed them for the security of ourselves and our nation. Nuclear weapons demonstrate the incredible scope of infrastructures for supporting national ambitions and nationalist identities.

In addition, and at a more local level, everyday encounters with roads, bridges, and other infrastructures that represent our national affiliations continually reinforce these connections and provide a daily reaffirmation of nationalist identities. Anthropologist Michael Billig describes these everyday encounters with symbols of national identity as "banal nationalism," such as when a citizen walks past a flag hanging from a government office building.[41] In this view, nationalism is not an exceptional outburst of patriotism but instead an ongoing identity in need of maintenance, continually flagged through the signs and symbols we encounter each day, including our interactions with the infrastructural state. Banal nationalism represents a way that infrastructures can themselves repair collective identities and provide an often subtle yet pervasive means by which "a nation must be put to daily use" through its infrastructures.[42]

With that said, we also have to emphasize the limits of nationalism and its power to legitimate infrastructural projects. Some feel the pull of nationalism more strongly than others, and some actively resist nationalist projects centered in infrastructures; the antinuclear power and weapons movements are good examples of ideologies and organizations counter to the broader sense of a shared national identity.[43] And some people and groups, marginalized through their race, ethnicity, religion, or other forms of difference, are simply left out of the collective stories that undergird nationalism.

A Tale of Two Canals: Infrastructures, Nationalism, and Decolonization

National interests and identities that are built into infrastructures can be a site for conflicting ideologies. Anthropologist Ashley Carse's study of the construction of the Panama Canal emphasizes the interests that the US state had in its construction during the late nineteenth- and early twentieth-century period of American imperial expansion through the Western Hemisphere. When Panama asserted its independence from Colombia in 1903, US warships were present to ensure the security of the fledgling state and stake a literal land claim to the canal site. As the Panamanian state sought to assert control over the canal, however, it increasingly framed American control of the waterway as an assault on its national sovereignty.[1]

Benedict Anderson notes that the colonial powers effectively taught the colonized how to think through and enact their own nationalist identities: "For the paradox of imperial official nationalism was that it inevitably brought what were increasingly thought of and written about as European 'national histories' into the consciousnesses of the colonized."[2] One after another, colonized peoples used nationalism to promote their own interests and attempt to wrest control of infrastructures away from the imperial powers. For example, when Egyptian president Abdel Nasser nationalized the Suez Canal in 1956, Israeli, French, and British forces invaded Egypt and attempted to reestablish control of the waterway, with the British especially fearful of losing strategic access to this shipping lane. Egypt maintained control of the Suez Canal, signaling for many the relative decline in power of the once global British Empire.[3] Similarly, Panama and the United States signed a treaty in 1977, in which the United States promised to cede control of the Panama Canal to Panama, which was effected on December 31, 1999.

1 Ashley Carse, *Beyond the Big Ditch: Politics, Ecology, and Infrastructure at the Panama Canal* (Cambridge, MA: MIT Press, 2014), 45–47.
2 Benedict Anderson, *Imagined Communities: Reflections on the Origin and Spread of Nationalism* (London: Verso, 1983), 120.
3 David M. McCourt, *Britain and World Power since 1945: Constructing a Nation's Role in International Politics* (Ann Arbor: University of Michigan Press, 2015), chap. 2.

The second broad mechanism we treat in this section centers on nations' infrastructural demonstrations that are meant to be anything but banal. Anthropologist James Scott, in *Seeing Like a State*, provides several examples of how states use building projects and infrastructural demonstrations as sociotechnical symbols of their power and prestige.[44] Scott emphasizes that many of these projects employed expert systems of knowledge that valorized theoretical and political ideals over systems of knowledge in place in local communities. Whereas the French engineers in Mukerji's history of the Canal du Midi appropriated local knowledge systems to solve engineering challenges, Scott notes a disjuncture between national ideology and local knowledge, often layered with class- and race-based prejudices, where experts overlook the needs of local and indigenous communities and dismiss their potential contributions. Scott discusses the case of the city of Brasília, founded in 1960 to replace Rio de Janeiro as the new capital of Brazil.[45] Brasília was built as a planned city according to the ideals of twentieth-century architects such as Le Corbusier and Oscar Niemeyer. Le Corbusier, in particular, abhorred the tangle of streets, alleyways, and communal spaces that typified older cities, such as London and Paris, as they grew organically over centuries. Brasília, by contrast, was designed on a grid structure and was meant to resemble an aircraft when viewed from above—a technology associated with modern ideals of clean lines and the highest technological sophistication.

The daily rhythms of residential life and work in Brasília were subordinate to the modern aesthetic, and Scott emphasizes that plans for the city "made not the slightest concession to the desires, history, and practices of its residents."[46] This top-down approach to infrastructural development is all too common as an expression of state power, yet we must also acknowledge how people and communities use infrastructures in creative and resistant ways to repair their own relationships and interests with respect to these omnipresent structures. Scott describes these everyday tools of resistance as the "weapons of the weak," where those subject to systems of power use passive and hidden forms of protest as a response to the state.[47] In the case of Brasília, historian Larissa Pires's work on the class, gender, and racial politics of the city demonstrates the limits of centralized planning for ordering urban spaces and their populations. Pires notes that the city's planners initially hoped that the migrant laborers who built the city would find other places to live upon its completion, but ultimately experts and elites had to relent and incorporate spaces for working-class settlers in Brasília (though only after a series of violent conflicts, including a massacre at a city labor camp in 1959).[48] The materiality of a city, however rationally planned, makes it hard to fully control according to the designs of politicians and planners, as Pires notes:

The fact that thousands of workers refused to return to their hometowns after finishing their initial construction assignments and, instead, fought for the right to reside in, and around, the city they had built shows how Brazil's state-controlled urban plans ultimately failed to control for the power of spontaneous urbanization. While architects, engineers, and technocrats planned the new city, its migrant population ultimately shaped its functions and dynamics; in other words, this demonstrates that experts can try to design a city from above, but it will be claimed from below.[49]

Specific components of city infrastructures can also be a key site for protest and resistance, as emphasized in the work of anthropologist Antina von Schnitzler, who writes on the role of infrastructures in postapartheid South Africa.[50] Von Schnitzler focuses on the implementation of utility meters as a means of regulating and commodifying water use in the transition to a postapartheid democracy. During the apartheid regime, many disenfranchised residents did not pay their rent and utility bills as form of protest against the state, limiting its ability to generate revenue. The continued use of nonpayment as a form of protest prompted the new regime to implement prepaid meters that would shut off utilities, especially water supplies, if residents did not pay in advance for service. In turn, residents sought ways to bypass or even destroy and remove the meters, along with demonstrations and other more visible forms of protest.

These two cases highlight the potential for infrastructures to serve as both mechanisms for state and elite influence over populations and resources for contesting power relations and structures. In both of these cases, as in the case of Barrio Logan and the Coronado Bridge, acts of resistance modified and appropriated elements of infrastructure from below, in defiance of the visions of the systems' original architects. In the South African case, this culminated in a protracted back-and-forth repair struggle between infrastructure owners and users, in which each side ultimately used physical modifications to infrastructure to try to limit the actions of the other. In cases like these, infrastructures' sociomateriality provides the ground for both repression and resistance, and repair becomes an explicitly political act.

Conclusion: Neoliberal Approaches to Infrastructural Repair in Twenty-First-Century Modernity

Although there is not a linear path from the gardens of Versailles that Louis XIV built in the late seventeenth century to the tools of Armageddon developed by the nuclear nation-states in the twentieth century, in this chapter we have emphasized a set of processes and stakeholders that were key factors in building the infrastructural state over the past three hundred years. State power has been a central theme, and states and state elites gain power through infrastructural systems in diverse ways, including through the economic capital derived from transportation networks, energy generation, real estate speculation, food production, and all the ways that land, water, and other environmental services can be tapped and commodified through infrastructures. Canals, roads, communications networks, and weapons systems also allow states to extend their administrative and military reach across large territories, controlling vast spaces and populaces. More subtly, though incredibly powerfully, nationalism and other internalized belief systems shape citizens' perspectives on their place in history, space, and everyday practices, meaning that states and other elite actors have the tools to set both the material and cultural frames for our lives.

Add this all up, and we have incredibly complex and pervasive sociotechnical systems that now stretch across the globe and give states enormous infrastructural power.[51] Elites did not do this work on their own, and in this chapter, we have also highlighted the work of experts and expert systems of knowledge that grew in tandem with the infrastructural state. In some cases, experts are elites in their own right; in others, they do not have formal credentials or state sponsorship, but instead have hard-won experiential knowledge of local settings that might be appropriated for state use or serve as tools for resistance to state interests. The very fact that infrastructures are embedded in our everyday lives means that they often serve as a site of contestation between elites, experts, and those who may be disenfranchised or subject to oppression and violence through these systems. Those struggles may be fought through the materiality of infrastructures via repair, such as when a water user in South Africa hacks the meter measuring their water use. Typically those material methods of resistance go hand-in-hand with discursive practices that seek to reframe infrastructural systems and shift the focus of repair from the maintenance that elites typically favor toward a more transformative vision.

As we prepare to consider more fully the global scale of infrastructural repair in the next chapter, we conclude here by discussing the continued role of infrastructures in shaping the political structures of modern nation-states. Scholars dispute the terms and concepts to understand where we stand in terms of our current relationship to modernity, with some arguing that we have never really been modern, that we are in a period of reflexive modernity, or that we reside now in postmodernity. In terms of the role of infrastructures, however, one thing is clear: we live in a world with an incredible number of these systems, entangled in every possible way with our material and political lives, at scales from local to global.[52] As more infrastructures are embedded within state systems and practices and older ones need to be repaired and replaced, we face critical questions about the state's role in supporting infrastructural systems—namely, who is responsible for building and maintaining all these infrastructures, and who pays? A fuller discussion of these questions requires backing up to the mid-twentieth century and tracing ideologies of state support for infrastructural development through the past several decades.

In the aftermath of the Great Depression and through the first two decades after World War II, the prevailing economic philosophy for industrialized nations was Keynesianism, named for the ideas of John Maynard Keynes, a British economist. By keeping unemployment and inflation low, economic growth steady, and limiting the duration of recessions, all through centralized economic planning and intervention, Keynesian economists and bankers advocated that central banks take a strong and active role in directly controlling their national economies. Infrastructures played a key part in the Keynesian approach, especially during hard economic downturns, where nations were advised to borrow money and use it to create jobs and stimulate the economy, typically through building public works such as roads, bridges, and dams. In this way, the continued building of the infrastructural state played a key role in the economic structure of modern nations in the twentieth century. By promising and promoting infrastructural projects, states attempted to both directly stimulate economic growth and lay the groundwork for future economic development through improved transportation networks, increased energy production, new communication lines, and other structures that facilitated economic exchange. In addition, after World War II, infrastructural repair was seen as critical for restoring the shattered communities and economies of postwar nations.[53]

The Keynesian approach generally held sway until the 1970s, when political unrest, oil shocks, and runaway interest and inflation rates challenged the toolbox that Keynesian bankers and economists used to influence and maintain the economy. *Neoliberalism* is the term that scholars and policy-makers use to define the global economic and political order since roughly 1970. It is an ideology that favors individual rights and freedoms, a limited role for the state, and the valorization of free markets. Neoliberalism is often set in contrast to the Keynesian approach, though it would be wrong to put them entirely at odds, especially given that nations in the twenty-first century still use many of the same Keynesian tools for manipulating economic trends. However, beginning in the late 1970s, under political leaders such as Margaret Thatcher in the United Kingdom and Ronald Reagan in the United States, a generation of economists trained to favor a neoliberal approach took control of central banks and global institutions such as the International Monetary Fund and the World Bank. The neoliberals advocated a smaller role for the state in the economy and called for reduced taxation rates, increased privatization of government-controlled industries, and attacks on union power in these industries—such as Reagan's standoff with US air traffic controllers and Thatcher's confrontation with coal miners in the United Kingdom, both in the early 1980s. In these conflicts, discourses about fairness, freedom, and control were informed by competing visions of the role of the state for infrastructures and their role in employment and labor power. Neoliberals argued that infrastructure would be more efficient and nimbler if taken away from the bureaucratic control of the state. Those who favored a more protectionist approach to industries such as the transportation and energy sectors targeted by Reagan and Thatcher suspected (and continue to suspect) a money and power grab for public resources by elite interests.[54]

Now, after several decades under the neoliberal policy regime, scholars are tracing the impacts of this approach on the infrastructural state, including especially the impacts of privatization and declining attention to infrastructural maintenance and repair. The case of postapartheid South Africa is one example of how private utility providers commodify materials essential for life, especially water, that states might well be expected to provide as a right of citizenship for their residents.[55] The infrastructural state even shapes our relationships at the most personal levels. Anthropologist Elana Shever traces the privatization of the formerly state-controlled oil industry in Argentina, noting that shifting resources from public to private not only has widespread macroeconomic and social effects but also affects the individuals and families whose lives are entangled with infrastructural systems, often over the course of generations.[56] When Argentina privatized its state-owned energy industry, managers and workers on the ground frequently conceptualized these changes in terms of familial relationships—personifying the company and its workers through kinship ties such as that between parents and children. When infrastructures are privatized, the impacts are felt at each scale that we treat in the chapters of this book, from this local and personal level of family ties, all the way to global perceptions about the stability (or not) of the infrastructural state. Workers sold a conception of state and corporate entities as family have to reframe and repair their understanding of the place of these entities in their lives; as globalization shifts jobs and corporate headquarters around the world, workers in many locations face this same challenge.

In the next chapter, we discuss this global scale of repair in more detail, but as a closing point here, it is useful to consider how infrastructural states were built on resources that will not be available in the same ways in coming decades and what that suggests for the stability of states. Anthropologists Timothy Mitchell and Dominic Boyer point to the importance of energy—itself an infrastructural product—in the creation of broader national infrastructures during the Keynesian and neoliberal eras, emphasizing the key role of petroleum in literally fueling growth in the decades following World War II.[57] As neoliberal reductions in public support of infrastructures take hold, however, Boyer notes that the "legacy of 'public infrastructure' has become rather threadbare, capturing a general sense of evaporating futurity in the medium of corroded pipes and broken concrete."[58] As we understand more and more about the coming impacts of climate change, Boyer's reference to the future of infrastructures raises questions about how and whether the existing nexus of infrastructure and energy use (especially via oil and gas) will continue to be a viable model in an age of rapidly changing climate. When enough people begin to see infrastructures as unreliable, risky, and in need of repair beyond our capacities, we engage in a collective form of "broken world thinking" that may ultimately call into question the legitimacy of the infrastructural state to repair itself.[59]

5 Confronting the Anthropocene: Reflexive Repair in an Age of Global Infrastructures

Why Aren't You a Farmer?

Have you ever stopped and asked yourself, "Why am I not a farmer?" Or maybe you are a farmer, and you wonder why you do not have more colleagues. While there are still areas of the world where farming is the predominant livelihood, especially in Asia, where China and India alone account for more than half of the world's farms, the process of industrialization led to a sharp decline in the overall proportion of farmers in most Western countries through the twentieth century.[1] The numbers are extreme in the case of US agriculture, where about 90 percent of the working population was engaged in farming in 1790, just after the establishment of the new American state. Fast-forward to the twenty-first century, when fewer than 2 percent of US workers are primarily employed in farming, a downward trend with profound implications.[2]

What does it take for a nation to move from a set of sociotechnical structures where nearly everyone is growing food to another set where only a small fraction are doing that work, all in just two hundred years? Questions like these motivate this concluding chapter, where we examine the global reach of infrastructures and the repair challenges humans face after having engineered sociotechnical systems across every region of the Earth. Your identity as a farmer (or not) plays a small but important part in a larger set of trends that have important consequences for the future of human life, the role of infrastructures in that future, and the choices that we make about repairing them. If that seems overstated, consider the role of sociotechnical systems in the story of why you are not a farmer and how the methods and structures for food production and consumption connect to global challenges such as food insecurity and climate change.

Urbanization, an increasingly complex division of labor, and technological innovation are the factors most commonly cited when explaining this set of changes. But as fewer farmers produced more and more food, those remaining worked in tandem with financiers, agricultural scientists, and consumers to develop new institutions and systems that formed the backbone of industrialized agriculture. Major crops like wheat, corn, and soy were standardized and became interchangeable commodities, processed into all kinds of packaged foods found in contemporary grocery stores. Consumers, for their part, largely delegated both food production and preparation to multinational food corporations and developed new tastes and dietary practices (or at least acquiesced to the new system).[3] In line with the themes we explored in the previous chapter, the infrastructural state also cultivated the growth of expert systems of knowledge that played a critical role in developing the industrial food system. In the United States, both the Department of Agriculture and the land-grant university system were established and grew in the scope of their influence through the second half of the nineteenth century and into the twentieth. Intended to support the development of the nation's research and development capacity around agricultural production, these institutions spurred the growth of new scientific fields devoted to animal and plant breeding, agricultural engineering, and the extension and application of this knowledge to specific farm communities.[4]

In sum, industrial food systems are yet another example of a complex infrastructural system: a tangle of people, technologies, and institutions, interconnected with other critical infrastructures, such as transportation and communication networks, and embedded in deep structures of power and difference. This set of infrastructures allows US farmers to produce food on a massive scale—contributing nearly $1 trillion to the nation's gross domestic product in 2015 and based in land use that accounts for roughly half of all US land area.[5] At the same time, important trade-offs are associated with this system. First, it prioritizes foods processed from the commodity crops that are grown on the largest scales and contributes to the negative health consequences associated with the Western diet, such as heart disease and diabetes. These diseases disproportionately affect communities of color and low-income consumers, who are more likely to live in one of the 15.6 million US households that the US Department of Agriculture has designated as food insecure, or lacking "consistent, dependable access to enough food for active, healthy living."[6]

Second, in addition to the impacts on human health and inequities in who gets to eat healthy and accessible food, there are enormous environmental consequences attributable to an infrastructure that uses land and resources on this scale. Water used for agricultural purposes or running off agricultural land after storms is the largest known cause of nonpoint water pollution in the United States.[7] US agriculture also has global environmental impacts, especially through its contributions to climate change. Nine percent of all US greenhouse gas emissions come directly from the agricultural sector in the form of livestock "emissions" and crop cultivation practices. These direct sources of greenhouse gases do not include the indirect yet significant emissions due to food-related transportation, processing, preparation, and waste. In turn, farmers around the world are already facing the impacts of climate change through more variable climate and weather, especially extreme droughts and storms.[8] As climate change presents increasingly severe impacts through the twenty-first century, the future of our food production and its associated infrastructures is called into question: Will this still-young century present humanity with a new age of famine and scarcity?

This question presents a new dimension to our discussion of infrastructural repair, where the global scope and impact of our infrastructures go beyond any one system or region. In the twenty-first century, humanity confronts a kind of infrastructural reckoning: our technical systems for mastery of social and material life have transformed the world, with vast use of energy and resources, changes to global climate, and loss of countless ecosystems and species. This trend has led some scholars to claim that we are living in a new age, the Anthropocene, where our infrastructures have impacts akin to the meteor strikes or tectonic shifts that have caused global-scale environmental change in past epochs. Placing human activity on the same geological scale as asteroids and volcanoes emphasizes the pervasiveness of our influence and the ongoing ways in which infrastructural repair has had a cumulative impact on our environs. As humans have built infrastructures, solved problems with them, advanced and defended interests through them, and created our current infrastructural world, we have largely favored repair strategies that maintain and preserve the material and discursive investments and assumptions embedded in those infrastructures—a strategy of repair as maintenance. As we have emphasized in the past chapters, transformative change is a harder and more contentious mode of repair, and in some cases, it calls for entirely new infrastructures and social arrangements. If we are at a point in human history when maintaining our existing infrastructures might lead to our very extinction, we face a paradox centered on repair: *Can we repair infrastructural repair itself?*[9]

Repairing repair, or what we term *reflexive repair* and define in more detail below, typifies the dilemmas facing humanity in the Anthropocene and is the central topic we address in this concluding chapter. What are the properties of infrastructures that span and reshape the world, and how do we understand the process of repair for systems of this scope? We examine these questions in the context of international efforts to understand and mitigate climate change and movements to develop new ways of thinking about sustainability and infrastructure. Sustainability presumes that some of the same tools that created global infrastructures and their consequences can now reorient technologies, institutions, and discourses toward a greener form of modernity. Again, however, the interdependence of knowledge, practice, and global infrastructures points to the challenges of radically repairing the very systems that have created the conditions for the Anthropocene.

We also examine how repair on a global scale remains importantly connected to repair at other levels. It takes a lot of work to build and maintain infrastructures at this scale, yet this work retains a fundamentally local character. A focus on repair can therefore help us see how social and material systems connect the broadest and the most local scales of analysis. At the same time, examples such as computing networks and environmental monitoring systems show how knowledge about the world, and especially the impact of global changes in the Anthropocene, is tied to the infrastructures that help create and support these knowledge claims—such as environmental monitoring networks that keep track of climate readings around the world.[10] These connections further reinforce the paradox of repair in the Anthropocene: we cannot develop data and theories about "the world" without the very systems that are creating wide-scale change.

Repair, Reflexive Modernity, and Infrastructural Globalism

Our knowledge about the global impacts of infrastructure depends on the existence of global infrastructures, reflecting a self-referential or *reflexive* quality that is common to many sociotechnical systems. While *reflexivity* is

Locating the Anthropocene in Time and Space

Chemist Paul Crutzen and biologist Eugene Stoermer popularized the term *Anthropocene*, which has been more broadly adopted by a wide range of scholars over the past two decades.[1] Geologists propose and accept various segments of time in Earth's history by what they observe in the structure and content of sedimentation. When layers of rock and ice indicate a distinct shift in the composition of life, atmosphere, and climate on Earth, that evidence points to starting and ending dates of major events and very long-term trends in the history of the planet. These signature shifts in geological time are called "golden spikes," and the International Commission on Stratigraphy, which considers and ratifies evidence for the boundaries of geological time, places physical markers in locations around the world where a rock outcropping or ice sheet contains key evidence indicating a transition.[2] Where, either physically or metaphorically, would we drive the golden spike to mark the Anthropocene?

Crutzen, in a 2002 *Nature* article that helped to popularize the term *Anthropocene*, proposed "the late eighteenth century, when analyses of air trapped in polar ice showed the beginning of growing global concentrations of carbon dioxide and methane." Often termed "The Great Acceleration," this candidate for the golden spike emphasizes the impacts of carbon emissions since the Industrial Revolution.[3] Another argument suggests that the explosion of the first atomic bomb, in 1945, and subsequent nuclear bombings and tests, laid down a layer of radioactive fallout that will be evident in the sedimentary record for future researchers (human or otherwise).[4] A third school of thought reaches all the way back to the origins of human agricultural settlements, more than ten thousand years ago, as the time when large-scale transformation of land use and human-induced species and ecosystem loss extended the period of warming after the end of the Pleistocene.[5]

In each of these proposals, the Anthropocene was kick-started through new human sociotechnical systems. Because these systems have been developed over centuries or even millennia, and in widespread locations, debates about precisely where to locate the golden spike marking the start of the Anthropocene prove tricky. But for scientists who are focused on settling this question, one thing is clear: we cannot gather the knowledge needed to drive the Anthropocene's golden spike without more and complex infrastructures. Geographer Erle Ellis and colleagues make this point strongly in an article on the organizational and technological requirements for dating the Anthropocene: "Like the Anthropocene itself, building scientific understanding of the human role in shaping the biosphere requires both sustained effort and leveraging the most powerful social systems and technologies ever developed on this planet."[6]

Locating the Anthropocene in Time and Space (continued)

Infrastructures have helped create the conditions for the Anthropocene, but they also provide a lot of information about how, where, and when anthropogenic global change occurs, raising complex questions about the role of infrastructures and repair in both building and understanding the sociotechnical context for this age.

---

1 Simon L. Lewis and Mark A. Maslin, *The Human Planet: How We Created the Anthropocene* (New Haven: Yale University Press, 2018); Paul J. Crutzen, "Geology of Mankind," *Nature* 415 (January 3, 2002): 23.
2 International Commission on Stratigraphy, "ICS—GSSPs," accessed December 13, 2018, at http://www.stratigraphy.org/index.php/ics-gssps; Stanley C. Finney and Asier Hilario, "GSSPs as International Geostandards and as Global Geoheritage," in *Geoheritage*, ed. Emmanuel Reynard and José Brilha (Amsterdam: Elsevier, 2018), 179–189; Lewis and Maslin, *The Human Planet*.
3 Crutzen, "Geology of Mankind," 23; Will Steffen, Wendy Broadgate, Lisa Deutsch, Owen Gaffney, and Cornelia Ludwig, "The Trajectory of the Anthropocene: The Great Acceleration," *Anthropocene Review* 2, no. 1 (2015): 81–98. Donna Haraway discusses the merits of alternative terms to *Anthropocene*, such as *Capitolocene* and *Chthulucene*. See Donna J. Haraway, *Staying with the Trouble: Making Kin in the Chthulucene* (Chapel Hill, NC: Duke University Press, 2016), 47–51.
4 Jan Zalasiewicz, Colin N. Waters, Mark Williams, Anthony D. Barnosky, Alejandro Cearreta, Paul Crutzen, Erle Ellis et al., "When Did the Anthropocene Begin? A Mid-Twentieth Century Boundary Level Is Stratigraphically Optimal," *Quaternary International* 383 (2015): 196–203; Colin N. Waters, Jan Zalasiewicz, Colin Summerhayes, Anthony D. Barnosky, Clément Poirier, Agnieszka Gałuszka, Alejandro Cearreta, et al., "The Anthropocene Is Functionally and Stratigraphically Distinct from the Holocene," *Science* 351, no. 6269 (2016): 137.
5 Erle C. Ellis and Navin Ramankutty, "Putting People in the Map: Anthropogenic Biomes of the World," *Frontiers in Ecology and the Environment* 6, no. 8 (2008): 439–447.
6 Erle C. Ellis, Dorian Q. Fuller, Jed O. Kaplan, and Wayne G. Lutters, "Dating the Anthropocene: Towards an Empirical Global History of Human Transformation of the Terrestrial Biosphere," *Elementa: Science of the Anthropocene* 1 (2013): 1.

a contested term with many meanings,[11] it is relevant here in two senses. First, infrastructure systems are reflexive in a causal sense, because they often operate at scales where their impacts on the world play a major role in shaping their own operating environments. For example, they may end up depleting the very resources they need to function or changing consumer behavior in ways that create new demand for infrastructure services.

This leads to feedback loops that can stabilize or destabilize these systems. Second, this causal reflexivity requires reflexive, self-aware thinking from infrastructure engineers, operators, and policymakers: to protect against potentially destabilizing feedback loops, they need to have a well-tuned awareness of the limitations of their knowledge and the potential for unanticipated complexity in even the most carefully designed systems. The increasing need for this reflexive perspective has important implications for the design, repair, and maintenance of sociotechnical systems, and it is closely related to the "broken world thinking" that Steven Jackson sees as driving the current wave of interest in repair.[12] Historically, the emergence of these ways of thinking is tied to a broader crisis of confidence in the modern world, which calls into question many of the original ideals of modernity.

Sociologist Ulrich Beck connects this crisis of confidence with the emergence of what he calls the *risk society*, a global situation that is increasingly dominated by the problem of managing the risks of industrialization rather than its economic benefits.[13] In this new world, Beck argues, social inequalities are increasingly related to the distribution of risk exposure in the population rather than the distribution of the economic benefits of modernization. The risks of industrialization are themselves increasingly complex, difficult to define, and global in their impacts. Beck is particularly concerned with industrial accidents and pollution, using the 1986 Chernobyl nuclear accident and the consequent spread of radioactive contamination across national borders as a key example.[14] Along with sociologist Anthony Giddens and others, Beck describes this new situation, in which the modern world is increasingly focused on problems created by modernity itself, as a new phase of *reflexive modernity*.[15] Others argue that this is less a new phase of modernity than a reemergence of tensions that have always plagued human relationships with nature and technology.[16] Either way, the result has been an increasing global recognition of the limits, impacts, and fragility of modern sociotechnical systems.

Global infrastructures have arguably played a crucial role in the development of reflexive modernity, although this is not a major theme in the work of Beck, Giddens, or others who write about the risk society. Infrastructures have contributed both by creating global risks and enabling global knowledge of those risks. STS scholar Paul Edwards captures this dual impact of infrastructure through the concept of *infrastructural globalism*, the "phenomenon by which 'the world' as a whole is produced and maintained—as both object of knowledge and unified arena of human action—through global infrastructures."[17] In particular, Edwards shows how our ability to conceive of and monitor weather and climate as global phenomena depends on a huge amount of infrastructural work. It requires not just weather stations around the world and communications infrastructures to relay data, but also transnational institutions and scientific networks to work out common concepts and data standards.[18] Before this sociotechnical infrastructure existed, there was little understanding of the evolution of weather systems over time or how they moved across the globe, which made it difficult to conceive of weather or climate as posing global risks. Similar stories of sociotechnical system building can be told about almost all of the global infrastructure systems we rely on today, from the internet to food distribution systems.

Infrastructural globalism is key to understanding reflexive modernity and the risk society for several reasons. Most congruent with Beck's arguments, the development of global infrastructures is part and parcel of the process of industrialization, and infrastructure systems are responsible for many of the pollution risks Beck focuses on. However, at least two other aspects of infrastructural globalism also play a crucial role in the risk society. First, as Edwards argues, global infrastructures and their associated networks of institutions and expertise are prerequisites for any systematic notion of global risk; for example, it is only through global information systems, infrastructures, and organizations dedicated to environmental monitoring that we have any sense of the global risk posed by carbon emissions or nuclear contamination. Second, infrastructures not only create new risks; they place more humans and more complex equipment in new relationships to existing natural hazards and thus in harm's way. This creates new kinds of human vulnerabilities to weather, climate, and geological dangers, quite different from those that preindustrial societies faced. For example, an earthquake might affect a preindustrial community largely through the direct impact of landslides or tsunamis on the human body, but in a modern city, we have to worry about being injured by collapsing bridges or buildings or being cut off from infrastructure networks that connect us to distant sources of crucial supplies like food, water, and medicine. The increasing global spread of infrastructures makes these vulnerabilities a key part of the global risk society.

The hard questions reflexive modernity has raised about sociotechnical systems have led to new ways of understanding and managing those systems. Societal concern about the limits and unintended consequences of sociotechnical systems means that planning and development efforts increasingly focus on repairing existing systems to address limitations and manage unintended consequences. And when new systems are planned, more and more effort goes into anticipating how they could go wrong and building in provisions for mitigation and repair in advance.[19] As a result, concepts like sustainability and resilience are becoming increasingly important in design, planning, and engineering. Overall, these are positive developments toward developing a more productive and manageable relationship between technology and the natural world. This suggests that perhaps we should be asking some of the same potentially productive questions about repair itself.

Repair is certainly not alien to modernity. None of the vast infrastructures that characterize the modern world could have the effect they do without ongoing work to maintain them. By keeping these systems running smoothly and serving human needs, repair—particularly in the mode of repair as maintenance—may also contribute to many of the negative impacts of modernity. Even efforts to transform these systems to make them more resilient or sustainable may have unintended consequences by making them more desirable to use. This does not mean that repair is necessarily a bad thing or that its practitioners should not be recognized and rewarded. Rather, it is the reason that we need a concept of reflexive repair—that is, an approach to repair that considers and plans for the limitations and potential unintended consequences of repair itself. The analytical tools we have presented throughout this book are ultimately aimed at supporting this kind of practice and understanding.

Reflexive Repair and Sustainable Infrastructures: A University Campus Case Study

What does a sustainable infrastructure look like? Is it a contradiction in terms, given the place of infrastructural modernity in the Anthropocene? Sustainability is most often defined in the terms developed by the World Commission on Environment and Development (also commonly referred to as the Brundtland Commission) in 1987: meeting "the needs of the present without compromising the ability of future generations to meet their own needs."[20] This view of sustainability implies a form of reflexive repair, asking the current generation to engage in an honest assessment of their needs and desired quality of life, the resources and impacts associated with those needs, and how those choices affect the ability of future generations to live a comparable existence. Therefore, because infrastructures are such a core means of controlling and employing sociotechnical forces, and infrastructural globalism is centrally implicated in the challenge of the Anthropocene, sustainable infrastructures themselves must be built and repaired to ensure that the needs of the present and the future are balanced in their design and operation. Ideally, the future-oriented reflexivity of sustainability asks us to continually reflect on and repair our lives and the structures that support them and redirect our path away from practices and systems that are unsustainable and foreclose livable futures.

That is the ideal. Environmental studies scholars Kates, Parris, and Leiserowitz note that the Brundtland Commission's definition of sustainability is open to considerable "creative ambiguity" and the potential for a wide range of parties to each "project their interests, hopes, and aspirations onto the banner of sustainable development."[21] In addition, the role of those future generations is underspecified, and the definition could be taken to imply that the main inequities we should worry about are between today and tomorrow, potentially ignoring the already enormous disparities that exist in the present.[22] In the worst-case scenario, sustainability might be used as an umbrella to "greenwash" the very practices and interests that created the context for climate change and other global risks to develop and worsen.[23] However, Kates, Parris, and Leiserowitz also acknowledge that these ambiguities may open up opportunities for "diverse stakeholders and perspectives" to engage in a creative dialogue to facilitate "coordination of mutual action to achieve multiple values simultaneously and even synergistically" around the goal of sustainable development.[24]

Given these potential pitfalls and opportunities, how can we make sense of sustainability as a way of theorizing and achieving reflexive repair? In the remainder of this section, we explore a case study centered on contemporary attempts to make colleges and universities sustainable. Postsecondary education provides a useful test case for sustainable repair efforts, in large part because many universities are making substantial efforts to green their campus operations. In fact, hundreds of institutions of postsecondary education around the world (though predominantly in the United States) have made commitments to international sustainability standards, such as the Association for the Advancement of Sustainability in Higher Education (AASHE) and its Sustainability Tracking, Assessment and Rating System (STARS). STARS provides accounting tools for calculating and accrediting levels of achievement in sustainability. Practically an industry in itself now, a whole tangle of consortiums, each with its own punchy acronyms, support institutions of higher education as they make commitments to reducing carbon emissions, incorporating local and sustainable food, and other key dimensions of campus sustainability.[25]

To consider the promise and challenges of sustainable infrastructures, we share here a case study from the school where Henke works, Colgate University, which opened a new fitness center in 2012.[26] The new center was a big improvement on the old one, which was getting old and, frankly, a bit gross. In addition to its gleaming newness, the fitness center had one additional feature to brag about: it was Colgate's first campus building certified through the US Green Building Council's LEED system of sustainable building practices. LEED certification is an international standard that grades new or renovated buildings according to several metrics, including energy and water efficiency, incorporation of green building materials, and sustainable landscaping around the building.[27] The new fitness center earned a gold rating, which means it is especially green and efficient according to LEED's standards, and Colgate has since committed to seeking LEED certification for all new building projects (see figure 5.1).[28]

According to a well-regarded and widely adopted set of standards, the fitness center is an example of sustainable building. At the same time, the physical footprint of Colgate's campus expanded by fifteen thousand square feet, with a commensurate expansion in energy use and other impacts on the sustainability profile of the institution as a whole. This example is just one building from one institution, but some of the trends and associated questions about sustainability as a form of reflexive repair are well illustrated in this example, especially when we consider Colgate as one among many institutions with commitments to sustainability. As noted above, hundreds of institutions, including Colgate, have signed on to commitments through standards such as AASHE's STARS and LEED. The rapid growth of these efforts points to the potential for higher education to serve

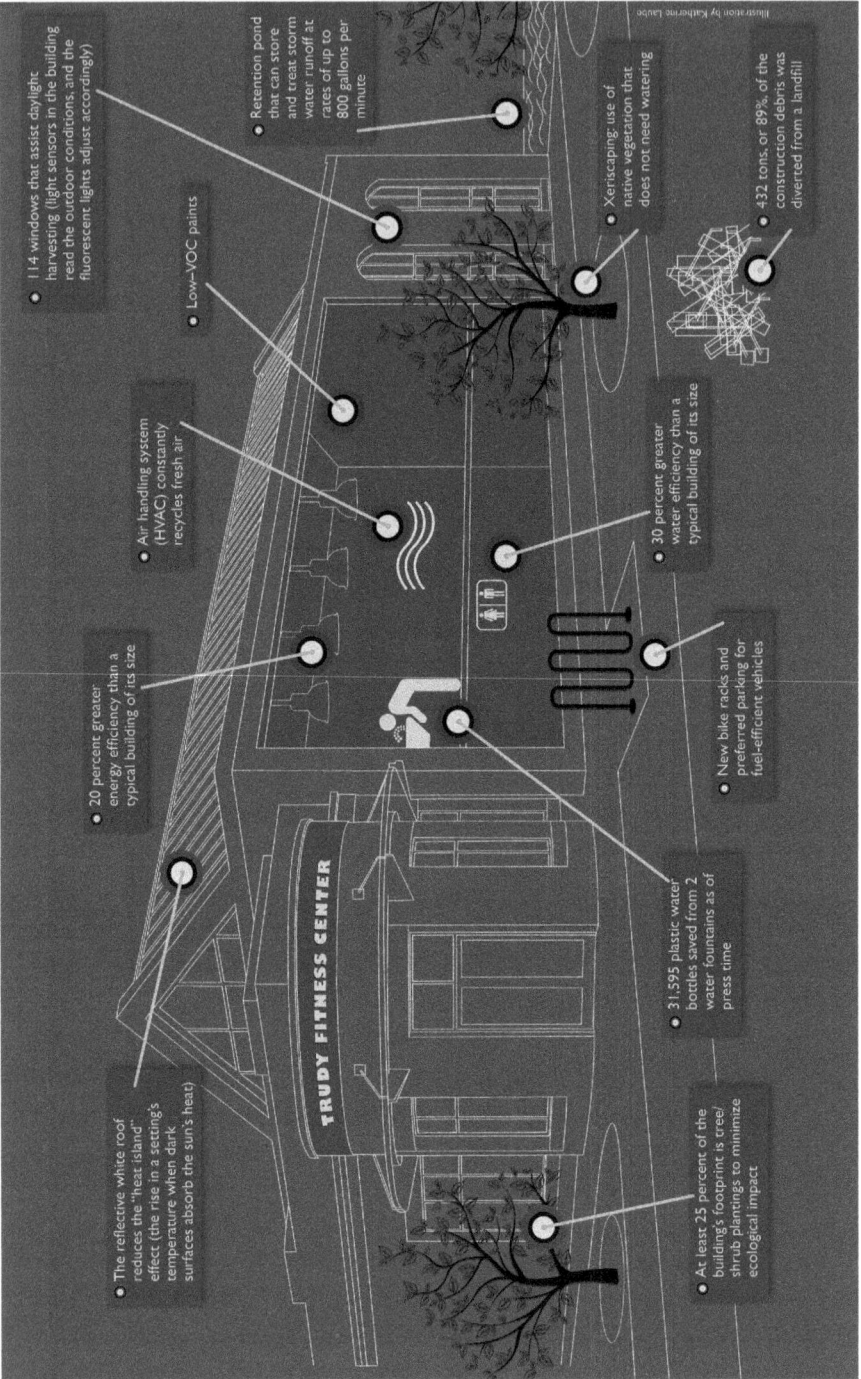

Figure 5.1

This illustration, included in a 2011 article for Colgate's alumni magazine, *Scene*, emphasizes the sustainable features of the fitness center, including elements that helped gain LEED certification for the building. Credit: Katherine Laube, Colgate University.

as a test sector for experimenting with sustainable infrastructures in relation to buildings and grounds, food procurement, travel, and other institutional features that contribute to, or are vulnerable to, climate change and
other environmental impacts. While often depicted as slow and stodgy in
adapting to new ideas and trends, universities can also be especially flexible
institutions given their relatively diffuse administrative structure and the
influence of new ideas and movements developed through research and
the continual generational turnover of students. Universities must balance
revenues and expenses, but the profit motive is less dominant than for
corporations, especially public corporations that must answer to investors
on a quarterly basis.

At the same time, it is important to consider why Colgate found it necessary to build a new fitness center. In US higher education, campus fitness
centers have acquired a kind of symbolic significance in recent years as universities increasingly compete to attract more and better applicants, even
while facing criticism over the growth in tuition costs as they add these
lavish amenities. The new fitness center certainly is an impressive stop on
a prospective student's tour of the campus, and there is a competitive pressure among similar institutions to keep up with the Joneses through new
facilities. In this way, universities tend to expand. Although the number of
students served might not change, the footprint of the physical campus, the
number of employees, and amenities provided may all increase, requiring
additional investment and growth. Andrew Ross terms universities a form
of "urban growth machine" for this reason, citing the case of New York
University as a prominent institutional real estate owner and developer in
Manhattan.[29] Higher education is big business and, in many communities,
a key source of employment, economic development, and gentrification.

If universities have the potential to be important sites for reflexive repair
centered on the goal of long-term sustainability—and yet are also key centers of economic growth—how can we use them to assess the potential
for change commensurate with the challenges of the Anthropocene? The
distinction between repair as maintenance and repair as transformation
can help us assess the scope of potential changes and to what extent those
new systems or practices maintain existing structures or reorient them
toward a more ambitious conception of sustainability. For example, the
easiest and most cost-effective changes for a university seeking to improve
its sustainability profile (and often ones that can save a lot of money) are

straightforward changes, such as replacing one technology or system with another that is more energy efficient. While a good starting point and an effort that we would certainly not want to discourage, this form of sustainability work is largely repair as maintenance: modest changes to operations may occur, but the essential things that a university does, and the broader political economy of higher education, largely have not changed.

This orientation toward maintenance is embedded in the standards developed by the organizations that accredit sustainability standards. AASHE's STARS system, for example, provides a comprehensive set of sustainability standards for colleges and universities, with indicators for nearly every aspect of campus operations, including carbon emissions, waste, and social sustainability.[30] These standards now essentially define what it means for a campus to be sustainable, providing a score and a symbolic medal (bronze, silver, gold, or platinum) to summarize progress toward green goals. Although helpful for planning sustainability initiatives and tracking progress, the standards may normalize a certain level of sustainability (or lack of sustainability) as being acceptable. Moreover, by making it possible to demonstrate commitment to sustainability by constructing new buildings, green building standards like those from LEED legitimate the continued growth of more elaborate college facilities. Taken as a whole, these elements may tend to encourage modest changes that essentially maintain the status quo.

It is not hard to understand why a standardized approach to sustainability is appealing. A repair-as-maintenance approach can be effective, especially in the short term and when directed toward the lower fruit of institutional change; this work may even result in net economic and environmental savings when energy use, water consumption, and waste are reduced. In addition, STS scholars Geoff Bowker and Susan Leigh Star, whose foundational work we reference in chapter 1, note that standards are increasingly built into the design and structure of our material environments, so perhaps the standardization of repair in specific infrastructural systems, including sustainability planning, should not be surprising.[31]

The maintenance approach to sustainability embedded in some sustainability standards stems in part from a focus on individual institutions, where a boundary is drawn around a specific university and that institution's local systems and infrastructures define the scope and rate of progress. Transformative repair likely means breaking through this boundary and planning sustainability efforts with a broader range of partners, including institutions outside postsecondary education. For example, when planning a new fitness center and other projects that are important to the sustainability of the campus, a university might consider ways to develop partnerships around planning for sustainability and resilience. A project that both supports a local initiative and makes a green impact on a regional or systemic infrastructure helps erase a false line drawn around a campus and encourages sustainability partnerships with an institution's neighborhood and regional communities. In this model of sustainability, the university as growth machine becomes harder to justify, especially as the key question shifts away from, "Is the fitness center LEED certified?" toward, "How does the new fitness center increase the sustainability and resiliency of our broader community?"[32] For instance, by investing in a local reforestation effort, a campus like Colgate could mitigate its carbon footprint and at the same time provide an amenity to support a range of community needs, including ecosystem restoration, protection from flooding and erosion, and the potential to develop new jobs and recreational opportunities.[33] Working cooperatively with other local institutions may also allow colleges and universities with fewer resources to share the costs of this work and address the inequities and injustices that are built into existing infrastructural systems, ideally supporting the long-term success of sustainability efforts for a broader set of participants.[34]

50,000 Metric Tons: A Ground-Level View of Sustainability and Reflexive Repair (Christopher R. Henke)

About 50,000 metric tons: that is our best estimate of how much carbon dioxide and other greenhouse gases my small, rural village contributes annually to global climate change. Hamilton is a college town in central New York State, site of Colgate University, and my home for the past twenty years. Our impact on climate change is small, considering the vast and rapidly warming Earth. And yet in an age of infrastructural globalism, everyone gets pulled into the politics of repair through practices and decisions in everyday life, and I am no exception. I drive a car (and get the oil changed at the repair shop), use a cell phone, and pay taxes to support construction and maintenance of roads, bridges, and many other forms of infrastructure. I contribute my share (or more) of carbon to the atmosphere through my daily interactions with infrastructures, and my community does too. And the impacts of a changing climate are already being felt here, with stronger storms, more extreme flooding

50,000 Metric Tons: A Ground-Level View of Sustainability and Reflexive Repair (continued)

events, and other climate challenges for the agricultural economy at the heart of this region.

The scope of the problem often seems overwhelming. How can one community address a problem so broad in its sources and implications? Over the past decade, I have become increasingly interested in ways to support my university and community in attempts to address the problem of climate change, working with a group of colleagues to form the Hamilton Climate Preparedness Working Group in 2016. The group brings together elected officials, community members, university staff, faculty and students, and others to reduce those 50,000 tons of carbon and prepare for a changing climate and its likely impacts on our region. We never discuss this work in terms of repair and the concepts used in this book, but I often reflect on the practical and existential challenges of making positive strides at the local and systemic levels, where even repair as maintenance is daunting, let alone a more fully transformative repair.

To chip away at those 50,000 tons, our preparedness group talks a lot about infrastructures. Gas lines, light bulbs, recycling bins, charging stations for electric vehicles, and bike lanes are just some of the infrastructural elements central to our conversations and plans. Though I sometimes find all the talking a bit frustrating, I remember at moments like these that infrastructural repair is not just about fixing things but also relationships and negotiation. An approach to reflexive repair that focuses only on technical fixes and eschews conversation misses opportunities to build common discourses and identities around a complex and urgent problem; talk also allows us to listen and learn when we disagree and misunderstand. Terms that are important for our analysis in this book—such as *infrastructural elites, experts*, or the *disenfranchised*, highlighted in the previous chapter—become more complex yet tangible when sitting in a meeting and listening to the village mayor, the head of municipal utilities, and community representatives. Indeed, given that considerations of equity and social justice often receive short shrift in climate plans and analysis, communities that do not build these conversations across lines of economic and social difference likely leave out those who may be most vulnerable to the impacts of climate change.[1]

---

1 Magnus Boström, "A Missing Pillar? Challenges in Theorizing and Practicing Social Sustainability," *Sustainability: Science, Practice, and Policy* 8, no. 1 (2012): 3–14; Chandra Russo and Andrew Pattison, "The Pitfalls and Promises of Climate Action Plans: Transformative Resilience Strategy in U.S. Cities," in *Resilience, Environmental Justice and the City*, ed. Beth Schaefer Caniglia, Manuel Vallée, and Beatrice Frank (New York: Routledge, 2017).

Conclusion: Infrastructural Politics and a Tool Kit for Reflexive Repair

As we draft this final section of the book, infrastructure is a frequent topic in the news. Current US president Donald Trump made infrastructure a key part of his 2016 election campaign, with promises of a spending package to help create jobs and stimulate the economy; a recent US government shutdown was centered around a disputed wall that Trump wants to build along the US-Mexico border. On the Left, there are calls for a Green New Deal, also meant to stimulate economic growth, but eschewing carbon-centered sources of development and focusing on sustainable technology. China, with its rapidly growing economy, dedicates almost 50 percent of its investment funding to infrastructure, including a significant amount on local projects throughout the country. In addition, China's Belt and Road initiative seeks to develop transportation infrastructures connecting Asia, Africa, and Europe.[35] Despite the ideological differences between these actors and states, they agree on one thing: infrastructures and infrastructural repair are the answer to just about every problem or question.

Another set of voices, much less invested in a future centered around current structures, critiques infrastructural modernity and its basis in a material globalism ever hungry for resources and unsustainable in its current form. Advocates of "degrowth" argue that global capitalism's never-ending search for new profit centers drives unsustainable rates of production and consumption and that this growth imperative must be abandoned for smaller and less consumption-centered forms of human organization.[36] The degrowth perspective points squarely at our dependence on a carbon-based economy to build contemporary structures that facilitate growth. Economist Giorgio Kallis notes the "free bonanza of work from fossil fuels" that allowed for "a monumental transformation of environments," including, of course, the creation of infrastructural globalism.[37]

While degrowth advocates argue that humans can reorient our lives toward a smaller impact and enjoy a simpler and less stressful lifestyle in the process, other critics of human infrastructures go further. The Dark Mountain Project, for example, a collective of writers and artists based largely in the United Kingdom, emphasizes the crisis at the center of the Anthropocene and uses the imagery of a failing machine to describe the wheels metaphorically coming off current sociotechnical structures:

> The crumbling empire is the unassailable global economy, and the brave new
> world of consumer democracy being forged worldwide in its name. Upon the
> indestructibility of this edifice we have pinned the hopes of this latest phase
> of our civilisation. Now, its failure and fallibility exposed, the world's elites are
> scrabbling frantically to buoy up an economic machine which, for decades, they
> told us needed little restraint, for restraint would be its undoing. Uncountable
> sums of money are being funnelled upwards in order to prevent an uncontrolled
> explosion. The machine is stuttering and the engineers are in panic. They are
> wondering if perhaps they do not understand it as well as they imagined. They
> are wondering whether they are controlling it at all or whether, perhaps, it is
> controlling them.[38]

In this view, repair is a key part of the problem, as elites and experts "scrabble frantically" to maintain the machine. Dark Mountain Project activists call for a process of "uncivilization" to reject and replace modern ideas of progress, growth, and the methods (and perceived failures) of mainstream environmental movements.[39]

Each of these visions of infrastructural futurity presents a discourse of infrastructural repair—whether to build it up or tear it down, to use a wall to keep out an imagined and demonized other, or to break apart structures that facilitate destruction and waste. Proposals about what to do with our infrastructures are invested with political imagery and language that embed specific relations of power in those discourses—and our own analysis here is no exception, of course. In some ways, we are optimists about the potential for infrastructures to improve human lives and bring people together, but at the same time, we are very aware of how they have formed the sociotechnical backstage for some terrible events and trends in the Anthropocene. Our recognition of this tension between the positive and negative aspects of infrastructure extends to our analysis of repair. In this book, we have been critical in many cases of repair work, especially where it serves to reproduce or reinforce social inequalities or negative environmental impacts. At the same time, we appreciate the creative work of repair artists discussed earlier in the book, like Willie the mechanic, practitioners of the paper towel trick, muralists who appropriate bridge columns as their own, and other brave bricoleurs who craft materiality for their own ends, often in the face of daunting structures of power. While it is a mistake to romanticize repair,[40] we will need all the skill and creative energy that is embodied in their work to find fair and sustainable solutions to the current challenges facing humanity.

Given both the ubiquity and the challenges of repair in an age of infrastructural globalism, we follow repair scholar Steven Jackson in asking, "Is there anything inherently hopeful about acts of repair? If forms of hope practiced through repair can be quietist and conservative, can they also be critical and political?"[41] Or, put another way, and especially from the point of view of a repair practitioner: What does a critical, reflective, and ultimately hopeful vision of repair entail? We conclude the book with a set of tools for reflexively repairing infrastructures, drawing on the tool kit of concepts that we have developed and employed throughout the text.

Reflexivity

Reflexive repair is a reorientation of repair itself, toward an ethic that considers the consequences of infrastructural globalism through the complex interactions of power, scale, and time that are built into the sociotechnical structures of modernity. In this concluding section, we provide a sense of how the concepts in our repair tool kit can be put to use for a positive practice of reflexive repair. In other words, how does one actually engage in a reflexive process that considers the trade-offs and consequences of repair actions? Ideally, these tools can be used by a wide range of repair persons and might even influence professional codes of conduct and other mechanisms of codifying best (reflexive) repair practices. Reflexive repair can be employed by mechanics, plumbers, and nurses—the workers we most commonly associate with everyday repairs—but also engineers, planners, policy analysts, investors, politicians, and anyone else whose activities shape and repair our infrastructural lives. When we refer to repair persons or repair workers in this chapter, it is with this broad scope in mind.

We provide more detail on the practice of reflexive repair in the sections that follow, but at this point, it is important to emphasize that reflexivity is not a cure-all for the power imbalances, disenfranchisement, and violence that can be facilitated via infrastructures. It is easy for us to say, "Think about it," but that admonishment is not enough, as reflexive repair can just as easily be used by those who wish to exercise and preserve power as it can be used to critique those practices and structures.[42] Consider, for example, that the original builders of the Coronado Bridge seem to have been just as aware of the local configurations of power and influence around the bridge as their successors who had to retrofit the bridge (see chapter 3). There were evidently some efforts to reach out to the communities around the bridge, understand their interests, and take this into consideration in the design of the bridge. The problem is, having looked at Barrio Logan and carefully considered its place in the local power structure, the bridge builders came to the realization that Barrio Logan's interests could be safely ignored and bulldozed through the community anyway; after all, it was the path of least resistance. Their successors were no more or less reflexive in their understanding; rather, the situation had changed in ways where it was clear the retrofit could not be accomplished without extensive, self-aware engagement with the community on a basis of mutual respect.

When we advocate reflexivity, we do so with particular values in mind—things we care about, like maintaining the Earth as a viable place for all of us to live, and ensuring justice and fairness in the distribution of the costs and benefits of infrastructure systems. Those who do not share those values with us will perhaps not be moved by these arguments (or will use them for other ends). However, there is a way in which the need for repair creates a space for reflection and an opportunity for considering alternate paths. Even in the most common forms of repair, such as a quickly corrected misunderstanding in a conversation between two people or a sharp kick to a sticking door, everyday hiccups in our relationships with infrastructures raise questions about why something did not work the way we expected and how to proceed. Furthermore, and to recall our earlier discussion of reflexive modernity, we are living in a time when bigger questions about infrastructural globalism are increasingly being asked. The need for repair does not beget shared values, but it does create a moment for questions and opportunities, suggesting at least the possibility for the kind of hopeful repair we cited above. For repair persons open to those possibilities, the other concepts in our repair tool kit provide more specific details about how to pose questions about infrastructural repair.

Expertise and the Challenges of Reflexive Repair: The Soil Quick Test

Experts keep appearing throughout this book, including in this chapter, and they seem likely candidates for exercising reflexive repair, given their centrality in defining the problems and promise of infrastructural systems. Despite their influence in both creating and diagnosing infrastructural globalism, however, experts often complain that no one listens to them. How much power do experts have to define the contours of global challenges like climate change, and how much can we depend on them to act as agents of repair for transformative change?

To address this question, we return to the case of agriculture and agricultural science. Following on the work and activism of figures like Rachel Carson, who raised alarms about the impact of pesticides on birds and other wildlife in the early 1960s, a new generation of scientists in the biological sciences paid increasing attention to the ecosystem-level impacts of chemicals, including for agricultural uses.[1] In a study of University of California farm advisors who work with farmers in California's fresh produce industry, Henke reported on the use of a soil test that allowed vegetable growers to check the level of nitrates present in their soil and make decisions about the need for additional fertilizers.[2] Farm advisors urged growers to make greater use of this "quick test" as a means of controlling the use of fertilizer and reducing the runoff of nitrates into local water supplies. Some communities near the vegetable farms had high levels of nitrate contamination in their drinking water, making it unsafe to drink.

However, despite its low cost and quick results, farmers were reluctant to adopt the quick test because fertilizer was relatively cheap and was the most important input for crop growth. As one farmer explained, "Nobody's gonna skip a $40 fertilizer application and possibly lose everything they got out there. They would rather make sure they have enough or too much. And it's hard to sell that on paper."[3] The advisors were frustrated, though they were used to this kind of response, given that they had no regulatory powers and depended only on the authority of their expertise and good personal relationships with the farmers. One advisor was reflective about the calculus that farmers used when considering the promise of the quick test for reducing fertilizer runoff: "It's less compelling for a farmer to make changes in fertilizer use [with the quick test method] than if I was to tell them they could change and increase production. . . . On the other hand, if the California [Environmental Protection Agency] hauls [the vegetable industry] into court . . . then I will suddenly be their best friend."[4]

The advisors' "hard sell" in this case was a suggested repair that clashed with farmers' values and practices for fertilizer use, revealing both the economic and legal power structure of the industry. In this case, the advisors were reflective and even a bit fatalistic about the politics of agricultural and environmental repair, quite aware of the limits of their ability to promote what seems (at least from the outside) a relatively straightforward repair to fertilizing practices. The farmers, for their part, were considering the quick test through calculations of relative economic risk and projections about future conditions, including legal and regulatory actions. Experts wield considerable power, but in the end, expert discourses are only as powerful as the structures that support them.

1 Scott Frickel, *Chemical Consequences: Environmental Mutagens, Scientist Activism, and the Rise of Genetic Toxicology* (New Brunswick, NJ: Rutgers University Press, 2004); Christopher R. Henke, "Changing Ecologies: Science and Environmental Politics in Agriculture," in *The New Political Sociology of Science: Institutions, Networks, and Power*, ed. Scott Frickel and Kelly Moore (Madison: University of Wisconsin Press, 2006), 215–243.
2 Henke, "Changing Ecologies"; Christopher R. Henke, *Cultivating Science, Harvesting Power: Science and Industry in California Agriculture* (Cambridge, MA: MIT Press, 2008).
3 Henke, "Changing Ecologies," 231.
4 Henke, "Changing Ecologies," 232.

Scale

Embedded in the idea of infrastructural globalism is the sheer complexity of infrastructural scale; as infrastructures increasingly form a single, interconnected global network, repair possibilities also begin to transcend individual infrastructure systems. The concept of reflexive repair suggests a need to understand the interconnections between multiple scales of repair, though the complexity of these interconnections can at times seem overwhelming. Consider the work of the Intergovernmental Panel on Climate Change (IPCC) as just one example of the mind-bogglingly complex trade-offs that have to be considered for repair at global scales. The 2014 IPCC report on climate mitigation runs to fourteen hundred pages, and that is just the report from one of three working groups.[43] Geared toward policymakers at the highest levels, the report includes analysis of the scale of infrastructures, from the project level to the nation, and suggests a range of regulatory, taxation, and investment strategies to reduce the carbon footprint associated with urban design and sprawl (see figure 5.2). The reality of infrastructural globalism means that a massive planetary repair manual on the scope of the IPCC reports is not just necessary, but is also, in fact, a remarkable achievement. However, from the point of view of a person or an institution facing the scope and scale of climate change, the IPCC is a very daunting resource for reflexive repair.

The connections and interdependencies of repair across scales call for a method that anticipates the way that infrastructures link together human actions and material structures. Repairs, even when they are aimed at a specific location, scale, or aspect of sociotechnical systems, tend to reverberate and have impacts at other levels because modern technologies and infrastructures are so integrated and interface with the environment in systematic ways. In fact, some repair efforts intended to promote sustainability or improve the efficiency of infrastructures, such as upgrading highways or switching to improved light bulbs, may paradoxically encourage behaviors that increase energy use and partially negate sustainability efforts, a phenomenon that environmental planners term the "rebound effect."[44]

With that understanding, we advocate two complementary approaches to help repair workers of all kinds think through the importance of scale. Each approach ties back to our discussion of infrastructural scale in prior chapters, where we detailed top-down and bottom-up approaches that

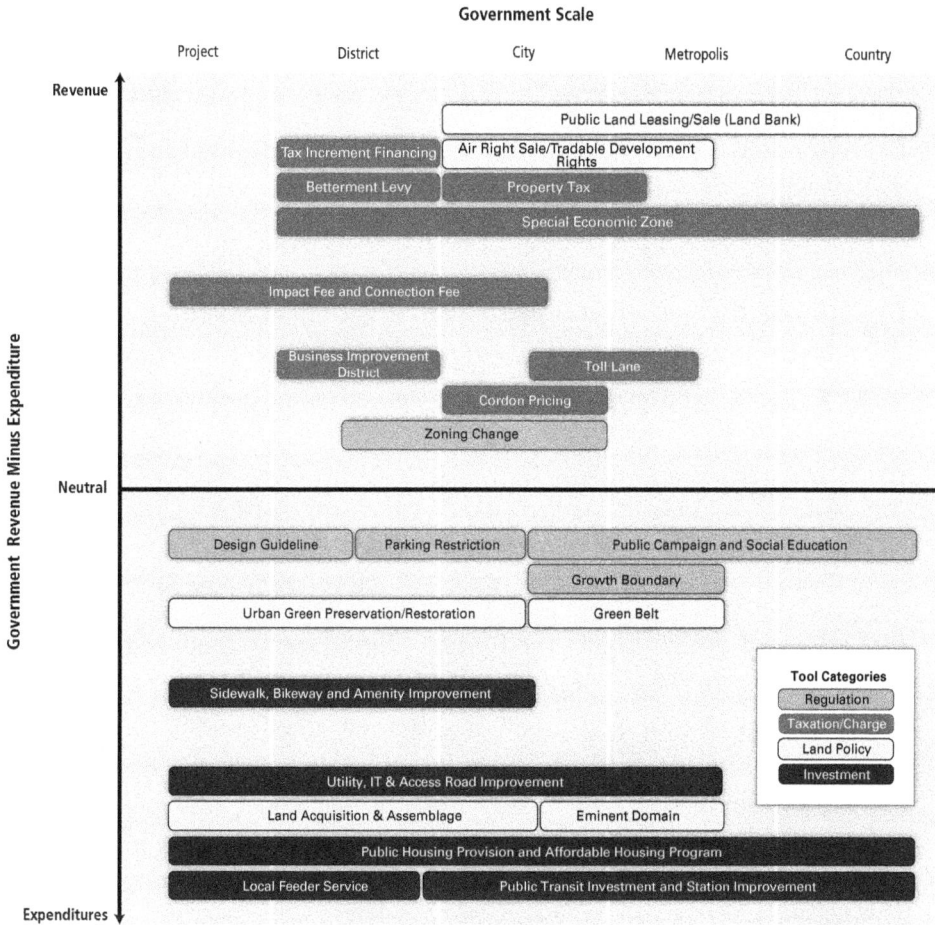

Government Scale

| Project | District | City | Metropolis | Country |

Revenue

- Public Land Leasing/Sale (Land Bank)
- Tax Increment Financing
- Air Right Sale/Tradable Development Rights
- Betterment Levy
- Property Tax
- Special Economic Zone
- Impact Fee and Connection Fee
- Business Improvement District
- Toll Lane
- Cordon Pricing
- Zoning Change

Neutral

- Design Guideline
- Parking Restriction
- Public Campaign and Social Education
- Growth Boundary
- Urban Green Preservation/Restoration
- Green Belt
- Sidewalk, Bikeway and Amenity Improvement

Tool Categories
- Regulation
- Taxation/Charge
- Land Policy
- Investment

- Utility, IT & Access Road Improvement
- Land Acquisition & Assemblage
- Eminent Domain
- Public Housing Provision and Affordable Housing Program
- Local Feeder Service
- Public Transit Investment and Station Improvement

Expenditures

Government Revenue Minus Expenditure

Figure 5.2

A chart from the 2014 IPCC report on climate change mitigation, graphically representing scales of government intervention (on the *x*-axis) and the resources that could fund those efforts (*y*-axis). Figure 12.20 from K. C. Seto, S. Dhakal, A. Bigio, H. Blanco, G. C. Delgado, D. Dewar, L. Huang, et al., *2014: Human Settlements, Infrastructure and Spatial Planning. In: Climate Change 2014: Mitigation of Climate Change. Contribution of Working Group III to the Fifth Assessment Report of the Intergovernmental Panel on Climate Change* (Cambridge: Cambridge University Press, 2014), 970.

scholars use in seeking to understand and explain how infrastructures operate as sociotechnical systems. Top-down approaches emphasize the structural roles of ecology, culture, politics, and markets, while the bottom-up perspective highlights the importance of everyday practices and material encounters with infrastructures. Brainstorming from the level of larger structures helps a repair person think through the broader implications of a repair, its potential side effects, and how it fits into larger systems of power and control. Seeing repair from the bottom-up perspective raises questions about how repair might affect the people who have to implement it or live with it on the ground level. Will it make their jobs harder? What kinds of behaviors is it likely to encourage or discourage? Will it reduce the quality of human or nonhuman life?

While the interconnectedness of infrastructural globalism, from local to global, means that all scales and perspectives matter, this bidirectional approach to considering the role of scale is a starting point that repair persons can use in a wide range of contexts. The two approaches provide a scalar lens for thinking through the people, institutions, and infrastructures that, linked together, create the superstructure for a phenomenon or problem. For example, in the case of climate change, those linkages could include an office worker putting a wet paper towel on a thermostat, a repair technician fixing a broken thermostat, a university administrator developing a plan to reduce the carbon footprint of the institution, a utility system engineer planning for changes to a power distribution system to account for new loads, a policy committee hashing out solutions to regional power interconnect problems, a government official's efforts to develop a policy to incentivize the use of renewable energy, and an international committee's recommendations on the comprehensive changes in technology and behavior needed to keep global warming below 2 degrees Celsius within the next century.

Delegates negotiating international climate accords are likely to focus first on this latter set of state and global scales and therefore may start by considering top-down views of repair. A committee facing up to the sustainability challenges of a university or other local or regional institution might consider how their work meshes (or not) with other institutions' practices and goals. Ideally, by alternating between top-down and bottom-up perspectives, repair persons can knit together multiple levels and begin to grasp the scalar complexity of repair.

Power and Invisibility, Discourse and Materiality

Infrastructures are incredibly powerful tools that enable many different human activities, especially through control of environmental conditions and forces, by reducing constraints on the human body and facilitating the compression of time and space. These properties of infrastructures allow humans to do some amazing things, and when those benefits and opportunities are widely available to most or all, infrastructures might be celebrated as an emancipatory force, a product of human ingenuity and labor that allows us to fulfill the dreams of the Enlightenment and its vision of steady human progress. However, we have shared many examples that highlight inequities in the benefits and capital that may accrue from infrastructures, and how repair is deployed to maintain those structures of inequality. Furthermore, with increasing recognition of the stark challenges presented by the Anthropocene, the impacts and externalities of infrastructural repair call into question the progressive connotations we might attach to the onset of infrastructural globalism and the possibilities it opens up for some portion of humanity.

The fact that infrastructures serve as global interfaces between human culture and nature means that their characteristics and impacts, whether we deem them good or bad, can sometimes be hard to see: the sociotechnical character of infrastructures naturalizes their structures, including imbalances in their power and effects. The expertise required to build and repair infrastructures is also difficult to attain and practice, and may seem almost like magic to those who do not have access to those skills and tools. Infrastructures are never truly invisible, but their salience to us, including their role in shaping power and privilege, shifts in and out of our consciousness based on their operation and to what extent we depend on or are oppressed by them. In addition, the discursive frames and meanings we attach to infrastructural forms provide the cultural context to understand and debate the shape and repair of infrastructures. This discursive work takes place alongside material forms of repair. Repair is not just a material, technological process, but rather a process that engages our core beliefs, assumptions, and ways of understanding each other and the world.

Reflexive repair in the context of infrastructural power, then, means asking questions that bring the sometimes obscure but always present properties of infrastructures to the surface, critically appraising the embedded dynamics of power, discourse, and materiality that are built into these sociotechnical structures. The distinction between repair as maintenance and as transformation is a key concern in the formation of such questions, especially: If repair is maintaining something, what exactly is being maintained? That broader question about maintenance can be broken down into a more targeted set of three questions, based in concepts from the repair tool kit. First, does a proposed repair largely reproduce existing power structures, or does it seek to remedy inequalities or injustices built into existing infrastructures? Second, does a proposed repair account for the people, communities, and ecologies most centrally affected by the material and discursive shape of an infrastructure? Who has a stake, and are they given a voice? Third, does a proposed repair solution restrict or close off possible futures, especially those we can imagine wanting for ourselves and future generations?

Good repair starts with questions that hypothesize about trouble and point toward solutions. By reflectively asking these questions about the practice of repair itself, we may develop better methods and tools to face the challenges of our infrastructural future.

Notes

Chapter 1

1. U.S. National Transportation Safety Board, "Collapse of the I-35W Highway Bridge, Minneapolis, Minnesota, August 1, 2007," November 14, 2008, xiii.

2. U.S. National Transportation Safety Board, "Collapse of the I-35W Highway Bridge," 14–15. Video of the collapse is available at http://minnesota.publicradio .org/collections/special/2007/bridge_collapse/video/ (accessed January 4, 2019).

3. U.S. National Transportation Safety Board, "Collapse of the I-35W Highway Bridge," xiii, 117–118; Paul Levy, "4 Dead, 79 Injured, 20 Missing after Dozens of Vehicles Plummet into River," *Minneapolis Star-Tribune*, August 2, 2007, http://www.startribune .com/4-dead-79-injured-20-missing-after-dozens-of-vehicles-plummet-into-river/ 11593606.

4. This includes one of the authors (Sims), who grew up in Minneapolis and undoubtedly drove over the bridge and even canoed under it numerous times, and remembers nothing about it.

5. Gray Plant Mooty, "Investigative Report to Joint Committee to Investigate the I-35W Bridge Collapse" (May 2008), 43, https://www.leg.state.mn.us/docs/2008/ other/080513/Investigative_Report.pdf.

6. Christopher R. Henke, "The Mechanics of Workplace Order: Toward a Sociology of Repair," *Berkeley Journal of Sociology* 44 (2000): 55–81. See also Stephen Graham and Nigel Thrift, "Out of Order: Understanding Repair and Maintenance," *Theory, Culture and Society* 24, no. 3 (2007): 1–25; Steven J. Jackson, "Rethinking Repair," in *Media Technologies: Essays on Communication, Materiality, and Society*, ed. Tarleton Gillespie, Pablo Boczkowski, and Kirsten Foot (Cambridge, MA: MIT Press, 2014), 221–239.

7. Oxford English Dictionary, "Repair, n.2," in *OED Online* (Oxford University Press, December 2018), http://www.oed.com/view/Entry/162629.

8. Emanuel A. Schegloff, Gail Jefferson, and Harvey Sacks, "The Preference for Self-Correction in the Organization of Repair for Conversation," *Language* 53, no. 2 (1977): 361–382; Emanuel A. Schegloff, "Repair after Next Turn: The Last Structurally Provided Defense of Intersubjectivity in Conversation," *American Journal of Sociology* 97, no. 5 (1992): 1295–1345; Emanuel A. Schegloff, "Third Turn Repair," in *Towards a Social Science of Language: Papers in Honor of William Labov*, ed. G. R. Guy, M. C. Feagin, D. Schiffrin, and J. Baugh (Amsterdam: John Benjamins, 1997), 31–40.

9. The widespread use of this term in science and technology studies goes back at least as far as the essays in Wiebe E. Bijker and John Law, eds., *Shaping Technology/ Building Society: Studies in Sociotechnical Change* (Cambridge, MA: MIT Press, 1992). It appears occasionally in the earlier set of essays in *The Social Construction of Technological Systems: New Directions in the Sociology and History of Technology*, ed. Wiebe E. Bijker, Thomas P. Hughes, and Trevor J. Pinch (Cambridge, MA: MIT Press, 1987).

10. Mooty, "Investigative Report," 22.

11. Mooty, "Investigative Report," 24.

12. Mooty, "Investigative Report," 25.

13. U.S. National Transportation Safety Board, "Collapse of the I-35W Highway Bridge," 49–50; Barry B. LePatner, *Too Big to Fall: America's Failing Infrastructure and the Way Forward* (New York: Foster Publishing, 2010), 9.

14. U.S. National Transportation Safety Board, "Collapse of the I-35W Highway Bridge," 48.

15. U.S. National Transportation Safety Board, "Collapse of the I-35W Highway Bridge," 56.

16. U.S. National Transportation Safety Board, "Collapse of the I-35W Highway Bridge," 53.

17. U.S. National Transportation Safety Board, "Collapse of the I-35W Highway Bridge," 54–55.

18. U.S. National Transportation Safety Board, "Collapse of the I-35W Highway Bridge," 60; LePatner, *Too Big to Fall*, 15–21.

19. U.S. National Transportation Safety Board, "Collapse of the I-35W Highway Bridge," 61.

20. U.S. National Transportation Safety Board, "Collapse of the I-35W Highway Bridge," 23; LePatner, *Too Big to Fall*, 5.

21. U.S. National Transportation Safety Board, "Collapse of the I-35W Highway Bridge," 150–151; LePatner, *Too Big to Fall*, 10.

22. U.S. National Transportation Safety Board, "Collapse of the I-35W Highway Bridge," 32, 63.

23. LePatner, *Too Big to Fall*.

24. David Montgomery, "Many Bridges Found Deficient after I-35W Collapse. Here's How Minnesota Responded," *Pioneer Press/TwinCities.Com*, July 30, 2017, https://www.twincities.com/2017/07/30/after-collapse-minnesota-fixed-deficient-bridges/; David Schaper, "10 Years after Bridge Collapse, America Is Still Crumbling," *National Public Radio*, August 1, 2017, https://www.npr.org/2017/08/01/540669701/10-years-after-bridge-collapse-america-is-still-crumbling.

25. U.S. National Transportation Safety Board, "Collapse of the I-35W Highway Bridge," 153–155.

26. Judith A. Martin, "Neighborhoods Confront a Disaster Aftermath," in *The City, the River, the Bridge: Before and after the Minneapolis Bridge Collapse*, ed. Patrick Nunnally (Minneapolis: University of Minnesota Press, 2011), 57–75.

27. See the timeline at https://www.leg.state.mn.us/lrl/guides/guides?issue=bridges (accessed January 4, 2019) and the account in Patrick Nunnally, "Building the New Bridge: Process and Politics in City-Building," in *The City, the River, the Bridge: Before and After the Minneapolis Bridge Collapse*, ed. Patrick Nunnally (Minneapolis: University of Minnesota Press, 2011), 35–54.

28. Quoted in William H. Batt, "Infrastructure: Etymology and Import," *Journal of Professional Issues in Engineering* 110, no. 1 (1984): 2. Also see Paul N. Edwards, "Infrastructure and Modernity: Force, Time, and Social Organization in the History of Sociotechnical Systems," in *Modernity and Technology*, ed. Thomas J. Misa, Philip Brey, and Andrew Feenberg (Cambridge, MA: MIT Press, 2003), 185–225.

29. Stephen J. Collier and Andrew Lakoff, "The Vulnerability of Vital Systems: How 'Critical Infrastructure' Became a Security Problem," in *The Politics of Securing the Homeland: Critical Infrastructure, Risk and Securitisation*, ed. Myriam Dunn and Kristian Soby Kristensen (London: Routledge, 2008).

30. Batt, "Infrastructure"; Ashley Carse, "Keyword: Infrastructure: How a Humble French Engineering Term Shaped the Modern World," in *Infrastructures and Social Complexity: A Companion*, ed. Penelope Harvey, Casper Bruun Jensen, and Atsuro Morita (London: Routledge, 2016), 27–39.

31. Batt, "Infrastructure"; Henry Petroski, *The Road Taken: The History and Future of America's Infrastructure* (New York: Bloomsbury, 2016), 11–20.

32. Petroski, *The Road Taken*, 20–27.

33. See the historical summary at https://www.infrastructurereportcard.org/making-the-grade/report-card-history/ (accessed January 4, 2019).

34. For example, Susan Saulny and Jennifer Steinhauer, "Bridge Collapse Revives Issue of Road Spending," *New York Times*, August 7, 2007, http://www.nytimes.com/2007/08/07/us/07highway.html.

35. Emma G. Fitzsimmons, "What Trump, Clinton and Voters Agreed On: Better Infrastructure," *New York Times*, November 9, 2016.

36. Collier and Lakoff, "The Vulnerability of Vital Systems."

37. Susan Leigh Star and Karen Ruhleder, "Steps toward an Ecology of Infrastructure: Design and Access for Large Information Spaces," *Information Systems Research* 7, no. 1 (1996): 111–134; Geoffrey C. Bowker and Susan Leigh Star, *Sorting Things Out: Classification and Its Consequences* (Cambridge, MA: MIT Press, 1999); Susan Leigh Star, "The Ethnography of Infrastructure," *American Behavioral Scientist* 43, no. 3 (1999): 377–391; Stephen Graham and Simon Marvin, *Splintering Urbanism: Networked Infrastructures, Technological Mobilities and the Urban Condition* (London: Routledge, 2001); Edwards, "Infrastructure and Modernity"; Paul N. Edwards, Steven J. Jackson, Geoffrey C. Bowker, and Cory P. Knobel, "Understanding Infrastructure: Dynamics, Tensions, and Design" (Washington, DC: National Science Foundation, January 2007), http://www.si.umich.edu/~pne/PDF/ui.pdf; Benjamin Sims, "Things Fall Apart: Disaster, Infrastructure, and Risk," *Social Studies of Science* 37, no. 1 (2007): 93–95; Collier and Lakoff, "The Vulnerability of Vital Systems"; Paul N. Edwards, Steven J. Jackson, Geoffrey C. Bowker, and Cory P. Knobel, "Introduction: An Agenda for Infrastructure Studies," *Journal of the Association for Information Systems* 10, no. 5 (2009): 364–374; David Ribes and Thomas A. Finholt, "The Long Now of Technology Infrastructure: Articulating Tensions in Development," *Journal of the Association for Information Systems* 10, no. 5 (2009): 375–398; Benjamin Sims, "Disoriented City: Infrastructure, Social Order, and the Police Response to Hurricane Katrina," in *Disrupted Cities: When Infrastructure Fails*, ed. Stephen Graham (Milton Park, UK: Routledge, 2010), 41–53; Brian Larkin, "The Politics and Poetics of Infrastructure," *Annual Review of Anthropology* 42, no. 1 (2013): 327–343; Ashley Carse, *Beyond the Big Ditch: Politics, Ecology, and Infrastructure at the Panama Canal* (Cambridge, MA: MIT Press, 2014); Kathryn Furlong, "STS beyond the 'Modern Infrastructure Ideal': Extending Theory by Engaging with Infrastructure Challenges in the South," *Technology in Society* 38 (2014): 139–147; Sebastián Ureta, *Assembling Policy: Transantiago, Human Devices, and the Dream of a World-Class Society* (Cambridge, MA: MIT Press, 2015); Carse, "Keyword: Infrastructure."

38. Edwards et al., "Understanding Infrastructure"; Steven J. Jackson, Paul N. Edwards, Geoffrey C. Bowker, and Cory P. Knobel, "Understanding Infrastructure: History, Heuristics and Cyberinfrastructure Policy," *First Monday* 12, no. 6 (2007), http://www.firstmonday.dk/ojs/index.php/fm/article/view/1904.

39. Thomas P. Hughes, *Networks of Power: Electrification in Western Society, 1880–1930* (Baltimore: Johns Hopkins University Press, 1983).

40. Edwards et al., "Understanding Infrastructure"; Jackson et al., "Understanding Infrastructure"; Bowker and Star, *Sorting Things Out*; Martha Lampland and Susan Leigh Star, eds., *Standards and Their Stories: How Quantifying, Classifying and Formalizing Practices Shape Everyday Life* (Ithaca, NY: Cornell University Press, 2008); Star and Ruhleder, "Steps toward an Ecology of Infrastructure"; Lawrence Busch, *Standards: Recipes for Reality* (Cambridge, MA: MIT Press, 2011).

41. For example, Bowker and Star, *Sorting Things Out*; Star and Ruhleder, "Steps toward an Ecology of Infrastructure."

42. Bowker and Star, *Sorting Things Out*, 34; Star and Ruhleder, "Steps toward an Ecology of Infrastructure."

43. Bowker and Star, *Sorting Things Out*, 35; Star and Ruhleder, "Steps toward an Ecology of Infrastructure," 113.

44. Henke, "The Mechanics of Workplace Order"; Christopher R. Henke, *Cultivating Science, Harvesting Power: Science and Industry in California Agriculture* (Cambridge, MA: MIT Press, 2008).

45. Graham and Thrift, "Out of Order."

46. Tim Dant, "The Work of Repair: Gesture, Emotion, and Sensual Knowledge," *Sociological Research Online* 15, no. 3 (2010): 97–118, http://www.socresonline.org.uk/15/3/7.html.

47. For example, Joshua A. Bell, Briel Kobak, Joel Kuipers, and Amanda Kemble, "The Materiality of Cell Phone Repair: Re-Making Commodities in Washington, DC," *Anthropological Quarterly* 91, no. 2 (2018): 603–633; Carse, *Beyond the Big Ditch*; Marisa Cohn, "Convivial Decay: Entangled Lifetimes in a Geriatric Infrastructure," in *Proceedings of the 19th ACM Conference on Computer-Supported Cooperative Work and Social Computing* (New York: ACM, 2016), 1511–1523; Jérôme Denis and David Pontille, "Maintenance Work and the Performativity of Urban Inscriptions: The Case of Paris Subway Signs," *Environment and Planning D: Society and Space* 32, no. 3 (2014): 404–416; Jérôme Denis and David Pontille, "Material Ordering and the Care of Things," *Science, Technology, and Human Values* 40, no. 3 (2015): 338–367; Lara Houston, "Unsettled Repair Tools: The 'Death' of the J.A.F. Box," (paper presented at the Maintainers Conference, Stevens Institute of Technology, Hoboken, NJ, 2016), 1–13, https://static1.squarespace.com/static/56a8e2fca12f446482d67a7a/t/570e9c8f01dbae9c3322fa7d/1460575376524/Maintainers-Lara-Houston.pdf; Lara Houston, Steven J. Jackson, Daniela K. Rosner, Syed Ishtiaque Ahmed, Meg Young, and Laewoo Kang, "Values in Repair," in *Proceedings of the 2016 CHI Conference on Human Factors in Computing Systems* (New York: ACM Press, 2016), 1403–1414; Lara Houston and Steven J. Jackson, "Caring for the 'Next Billion' Mobile Handsets: Opening Proprietary Closures through the Work of Repair," in *Proceedings of the Eighth International Conference on Information and Communication Technologies and Development* (New

York: ACM, 2016), 10:1–10:11; Steven J. Jackson, Alex Pompe, and Gabriel Krieshok, "Repair Worlds: Maintenance, Repair, and ICT for Development in Rural Namibia," in *Proceedings of the ACM 2012 Conference on Computer Supported Cooperative Work* (New York: ACM Press, 2012), 107; Daniela K. Rosner and Morgan Ames, "Designing for Repair? Infrastructures and Materialities of Breakdown," in *Proceedings of the 17th ACM Conference on Computer Supported Cooperative Work and Social Computing* (New York: ACM, 2014), 319–331; Benjamin Sims, "Seismic Shifts and Retrofits: Scale and Complexity in the Seismic Retrofit of California Bridges," in *Retrofitting Cities: Priorities, Governance and Experimentation*, ed. Mike Hodson and Simon Marvin (London: Routledge, 2016), 13–33; Benjamin Sims and Christopher R. Henke, "Repairing Credibility: Repositioning Nuclear Weapons Knowledge after the Cold War," *Social Studies of Science* 42, no. 3 (2012): 324–347; Anissa Tanweer, Brittany Fiore-Gartland, and Cecilia Aragon, "Impediment to Insight to Innovation: Understanding Data Assemblages through the Breakdown-Repair Process," *Information, Communication and Society* 19, no. 6 (2016): 736–752; Sebastián Ureta, "Normalizing Transantiago: On the Challenges (and Limits) of Repairing Infrastructures," *Social Studies of Science* 44, no. 3 (2014): 368–392. See also the papers collected in Lara Houston, Daniela K. Rosner, Steven J. Jackson, and Jamie Allen, eds., "R3pair Volume," Special issue, *Continent,* no. 6.1, 2017, http://continentcontinent.cc/index.php/continent/issue/view/27; Ignaz Strebel, Alain Bovet, and Philippe Sormani, eds., *Repair Work Ethnographies: Revisiting Breakdown, Relocating Materiality* (Singapore: Palgrave Macmillan, 2019).

48. Jackson, "Rethinking Repair."

49. Jackson, "Rethinking Repair," 221.

50. Jackson, "Rethinking Repair," 222.

51. Jackson, "Rethinking Repair."

52. See http://themaintainers.org/ (accessed January 4, 2019).

53. Andrew Russell and Lee Vinsel, "Let's Get Excited about Maintenance!" *New York Times*, July 22, 2017.

54. Andrew Russell and Lee Vinsel, "Hail the Maintainers," *Aeon*, 2016, https://aeon.co/essays/innovation-is-overvalued-maintenance-often-matters-more; David Edgerton, *The Shock of the Old: Technology and Global History since 1900* (New York: Oxford University Press, 2007).

55. Russell and Vinsel, "Hail the Maintainers."

56. See Sulfikar Amir, ed., *The Sociotechnical Constitution of Resilience: A New Perspective on Governing Risk and Disaster* (Singapore: Palgrave Macmillan, 2018).

57. Charles Perrow, *Normal Accidents: Living with High-Risk Technologies*, 2nd ed. (Princeton, NJ: Princeton University Press, 1999).

58. Ulrich Beck, *Risk Society: Towards a New Modernity* (London: Sage, 1992); Ulrich Beck, Anthony Giddens, and Scott Lash, *Reflexive Modernization: Politics, Tradition and Aesthetics in the Modern Social Order* (Cambridge, UK: Polity Press, 1994); Anthony Giddens, "Risk and Responsibility," *Modern Law Review* 62, no. 1 (1999): 1–10; Anthony Giddens, *Runaway World: How Globalization Is Reshaping Our Lives* (New York: Routledge, 2000).

59. Bruno Latour, *We Have Never Been Modern* (Cambridge, MA: Harvard University Press, 1993).

60. Schegloff et al., "The Preference for Self-Correction in the Organization of Repair for Conversation"; Schegloff, "Repair after Next Turn"; Schegloff, "Third Turn Repair."

61. Sims, "Seismic Shifts and Retrofits."

62. Larkin, "The Politics and Poetics of Infrastructure."

63. Star and Ruhleder, "Steps toward an Ecology of Infrastructure." Bowker uses *infrastructural inversion* to describe an analytical approach that seeks to foreground normally invisible aspects of infrastructure. We see little purpose in distinguishing the analyst's experience of this phenomenon from users' similar experiences during breakdown, repair, and maintenance, and prefer to use *infrastructural inversion* to cover both (see the text box in chapter 3). Bowker and Star, *Sorting Things Out*, 34–37; the term is originally from Geoffrey C. Bowker, *Science on the Run: Information Management and Industrial Geophysics at Schlumberger, 1920–1940* (Cambridge, MA: MIT Press, 1994).

64. Everett C. Hughes, "Work and the Self," in *Social Psychology at the Crossroads*, ed. John H. Rohrer and Muzafer Sherif (New York: Harper and Brothers, 1951), 313–323; Blake E. Ashford and Glen E. Kreiner, "'How Can You Do It?' Dirty Work and the Challenge of Constructing a Positive Identity," *Academy of Management Review* 24, no. 3 (1999): 413–434. On the invisibility of technicians generally, see Steven Shapin, "The Invisible Technician," *American Scientist* 77 (1989): 554–563; Steven Shapin, *A Social History of Truth: Civility and Science in Seventeenth-Century England* (Chicago: University of Chicago Press, 1994), 355–407.

65. See Mary Douglas, *Purity and Danger: An Analysis of Concepts of Pollution and Taboo* (London: Routledge & Kegan Paul, 1966).

66. On "infrastructural bypass," see Graham and Marvin, *Splintering Urbanism*, 168–177.

67. See Chandra Mukerji, *Territorial Ambitions and the Gardens of Versailles* (Cambridge: Cambridge University Press, 1997); Chandra Mukerji, *Impossible Engineering: Technology and Territoriality on the Canal Du Midi* (Princeton, NJ: Princeton University Press, 2009); Carse, *Beyond the Big Ditch*.

68. Larkin, "The Politics and Poetics of Infrastructure," 336.

69. Benjamin Sims and Christopher R. Henke, "Maintenance and Transformation in the U.S. Nuclear Weapons Complex," *IEEE Technology and Society Magazine* 27, no. 3 (2008): 32–38; Sims and Henke, "Repairing Credibility."

70. Edwards, "Infrastructure and Modernity," 192–194.

71. Edwards, "Infrastructure and Modernity," 193.

72. Julian Orr, *Talking about Machines: An Ethnography of a Modern Job* (Ithaca, NY: Cornell University Press, 1996); Henke, "The Mechanics of Workplace Order."

73. See the essays on Hurricane Katrina in a special section of the journal *Social Studies of Science* (February 2007), collected and introduced by Sims, including Christopher R. Henke, "Situation Normal? Repairing a Risky Ecology," *Social Studies of Science* 37, no. 1 (2007): 135–142; Sims, "Things Fall Apart"; Benjamin Sims, "'The Day after the Hurricane': Infrastructure, Order, and the New Orleans Police Department's Response to Hurricane Katrina," *Social Studies of Science* 37, no. 1 (2007): 111–118.

74. Sims and Henke, "Maintenance and Transformation in the U.S. Nuclear Weapons Complex"; Sims and Henke, "Repairing Credibility."

75. Christopher R. Henke, "The Sustainable University: Repair as Maintenance and Transformation," *Continent* 6, no. 1 (2017): 40–45.

76. E. Summerson Carr and Michael Lempert, *Scale: Discourse and Dimensions of Social Life* (Berkeley: University of California Press, 2016).

77. Benjamin Sims, "Layers of Abstraction and the Organization of Repair in High Performance Computing" (paper presented at the Society for Social Studies of Science Annual Meeting, New Orleans, LA, 2019).

78. Benjamin Sims, "Concrete Practices: Testing in an Earthquake-Engineering Laboratory," *Social Studies of Science* 29, no. 4 (1999): 483–518.

79. Edwards, "Infrastructure and Modernity."

80. Ureta, "Normalizing Transantiago."

81. Benjamin Sims, "Making Technological Timelines: Anticipatory Repair and Testing in High Performance Scientific Computing," *Continent* 6, no. 1 (2017): 81–84. On the temporality of repair, see also Cohn, "Convivial Decay"; Marisa Cohn, "'Lifetime Issues': Temporal Relations of Design and Maintenance," *Continent* 6, no. 1 (2017): 4–12; Brittany Fiore-Gartland, "Technological Residues," *Continent* 6, no. 1 (2017): 25–29; Lara Houston, "The Timeliness of Repair," *Continent* 6, no. 1 (2017): 51–55.

Chapter 2

1. Tim Dant, *Materiality and Society* (New York: Open University Press, 2005), 111; Paul N. Edwards, "Infrastructure and Modernity: Force, Time, and Social Organization in the History of Sociotechnical Systems," in *Modernity and Technology*, ed. Thomas J. Misa, Philip Brey, and Andrew Feenberg (Cambridge, MA: MIT Press, 2003), 185–225.

2. There are good reasons for this history, including the danger of explaining gender or racial inequalities as material and genetic differences. Similarly, as noted in work in environmental sociology, there is an inherent assumption of "human exceptionalism" in mainstream sociological analyses, where humans are treated as though they do not exist within a system of material resources and constraints. See William R. Catton, Jr., and Riley E. Dunlap, "Environmental Sociology: A New Paradigm," *American Sociologist* 13, no. 1 (1978): 41–49; Riley E. Dunlap, and William R. Catton, "Struggling with Human Exemptionalism: The Rise, Decline, and Revitalization of Environmental Sociology," *American Sociologist* 25, no. 1 (1994): 5–30. Even where the term *material culture* is used to describe the place of objects, the focus, as Dant, *Materiality and Society* notes, is more frequently on the symbolic frames that human culture uses to understand its stuff.

3. Dant, *Materiality and Society*, 111–115. Design and human factors researchers similarly use the term *affordances* to describe how material artifacts allow users to take certain actions while restricting others; see Don Norman, *The Design of Everyday Things: Revised and Expanded Edition* (New York: Basic Books, 2013), 10–23, 123–161.

4. Bruno Latour, *The Pasteurization of France* (Cambridge, MA: Harvard University Press, 1984); Bruno Latour, "Where Are the Missing Masses? The Sociology of a Few Mundane Artifacts," in *Shaping Technology/Building Society*, ed. Wiebe E. Bijker and John Law (Cambridge, MA: MIT Press, 1992), 225–258; Bruno Latour, *Pandora's Hope: Essays on the Reality of Science Studies* (Cambridge, MA: Harvard University Press, 1999); Michel Callon, "Some Elements of a Sociology of Translation: Domestication of the Scallops and the Fishermen of St. Brieuc Bay," in *Power, Action, and Belief: A New Sociology of Knowledge?* ed. John Law (London: Routledge, Kegan, and Paul, 1986), 196–233; John Law, "Technology and Heterogeneous Engineering: The Case of Portuguese Expansion," in *The Social Construction of Technological Systems: New Directions in the Sociology and History of Technology*, ed. Wiebe E. Bijker, Thomas P. Hughes, and Trevor J. Pinch (Cambridge, MA: MIT Press, 1987), 111–134; John Law, *Organizing Modernity* (Cambridge, MA: Blackwell, 1994); John Law, *Aircraft Stories: Decentering the Object in Technoscience* (Durham, NC: Duke University Press, 2002); Donna J. Haraway, *Simians, Cyborgs, and Women: The Reinvention of Nature* (New York: Routledge, 1991); Karen Barad, "Posthumanist Performativity: Toward an Understanding of How Matter Comes to Matter," *Signs: Journal of Women in Culture and Society* 28, no. 3 (2003): 801–831; Edwin Sayes, "Actor–Network Theory

and Methodology: Just What Does It Mean to Say That Nonhumans Have Agency?" *Social Studies of Science* 44, no. 1 (2014): 134–149.

5. Latour, "Where Are the Missing Masses?"

6. Latour, "Where Are the Missing Masses?"

7. Christopher R. Henke, "The Mechanics of Workplace Order: Toward a Sociology of Repair," *Berkeley Journal of Sociology* 44 (2000): 67–68.

8. Douglas Harper, *Working Knowledge: Skill and Community in a Small Shop* (Berkeley: University of California Press, 1987).

9. Harper, *Working Knowledge*, 121.

10. Harper, *Working Knowledge*, 126.

11. Henke, "The Mechanics of Workplace Order," 64.

12. Henke, "The Mechanics of Workplace Order," 63; Susan Leigh Star and Karen Ruhleder, "Steps toward an Ecology of Infrastructure: Design and Access for Large Information Spaces," *Information Systems Research* 7, no. 1 (1996): 113; Tim Dant, "The Work of Repair: Gesture, Emotion, and Sensual Knowledge," *Sociological Research Online* 15, no. 3 (2010): 97–118, http://www.socresonline.org.uk/15/3/7.html.

13. John Hannigan, *Environmental Sociology: A Social Constructionist Perspective*, 2nd ed. (New York: Routledge, 2006), 36.

14. Henke, "The Mechanics of Workplace Order," 65.

15. W. I. Thomas, *The Unadjusted Girl* (Boston: Little, Brown, 1923); W. I. Thomas, *The Child in America* (New York: Knopf, 1928); Peter L. Berger and Thomas Luckmann, *The Social Construction of Reality: A Treatise in the Sociology of Knowledge* (Harmondsworth: Penguin, 1971).

16. Henke, "The Mechanics of Workplace Order," 70.

17. Henke, "The Mechanics of Workplace Order," 65–66.

18. Boris Kingma and Wouter van Marken Lichtenbelt, "Energy Consumption in Buildings and Female Thermal Demand," *Nature Climate Change* 5 (2015): 1054–1056.

19. Julian E. Orr, *Talking about Machines: An Ethnography of a Modern Job* (Ithaca, NY: Cornell University Press, 1996).

20. Orr, *Talking about Machines*.

21. Orr, *Talking about Machines*, 143.

22. Mona Chalabi, "Dear Mona, How Many Flight Attendants Are Men?" *FiveThirtyEight* (blog), October 3, 2014, https://fivethirtyeight.com/datalab/dear-mona-how-many-flight-attendants-are-men/.

23. Arlie Russell Hochschild, *The Managed Heart: The Commercialization of Human Feeling* (Berkeley: University of California Press, 1983); Amy S. Wharton, "The Sociology of Emotional Labor," *Annual Review of Sociology* 35 (2009): 147–165.

24. Hochschild, *The Managed Heart*, 119.

25. Erving Goffman, "On Face-Work: An Analysis of Ritual Elements in Social Interaction," *Psychiatry* 18 (1955): 213–231; Erving Goffman, *The Presentation of the Self in Everyday Life* (New York: Doubleday, 1959); Erving Goffman, *Interaction Ritual: Essays on Face-to-Face Behavior* (New York: Pantheon, 1967).

26. María Puig de la Bellacasa, "Matters of Care in Technoscience: Assembling Neglected Things," *Social Studies of Science* 41, no. 1 (2011): 85–106; María Puig de la Bellacasa, "'Nothing Comes without Its World': Thinking with Care," *Sociological Review* 60, no. 2 (2012): 197–216; Annemarie Mol, Ingunn Moser, and Jeanette Pols, "Care: Putting Practice into Theory," in *Care in Practice: On Tinkering in Clinics, Homes, and Farms*, ed. Annemarie Mol, Ingunn Moser, and Jeanette Pols (Bielefeld, Germany: Transcript Verlag, 2010), 7–26; Lara Houston and Steven J. Jackson, "Caring for the 'Next Billion' Mobile Handsets: Opening Proprietary Closures through the Work of Repair," in *Proceedings of the Eighth International Conference on Information and Communication Technologies and Development* (New York: ACM, 2016), 10:1–10:11.

27. Evelyn Nakano Glenn, "Creating a Caring Society," *Contemporary Sociology* 29, no. 1 (2000): 84–94. See also Mary Douglas, *Purity and Danger: An Analysis of Concepts of Pollution and Taboo* (London: Routledge and Kegan Paul, 1966.). Duffy's analysis of labor statistics shows that in the United States, this work has been disproportionately done by women and persons of color. Mignon Duffy, "Doing the Dirty Work: Gender, Race, and Reproductive Labor in Historical Perspective," *Gender and Society* 21, no. 3 (2007): 313–336.

28. Mol et al., "Care."

29. Edwards, "Infrastructure and Modernity."

30. Adam Greenfield, "A Sociology of the Smartphone," *Longreads* (blog), June 13, 2017, https://longreads.com/2017/06/13/a-sociology-of-the-smartphone/.

31. Daniela K. Rosner and Morgan Ames, "Designing for Repair? Infrastructures and Materialities of Breakdown," in *Proceedings of the 17th ACM Conference on Computer Supported Cooperative Work and Social Computing* (New York: ACM, 2014), 319–331.

32. Daniela Rosner and Jonathan Bean, "Learning from IKEA Hacking: I'm Not One to Decoupage a Tabletop and Call It a Day," in *Proceedings of the SIGCHI Conference on Human Factors in Computing Systems* (New York: ACM, 2009), 419–422; Rosner and Ames, "Designing for Repair?"; Lara Houston, Steven J. Jackson, Daniela K. Rosner, Syed Ishtiaque Ahmed, Meg Young, and Laewoo Kang, "Values in Repair,"

in *Proceedings of the 2016 CHI Conference on Human Factors in Computing System* (New York: ACM Press, 2016), 1403–1414; Tim Dant, "Inside the Bicycle: Repair Knowledge for All," in *Repair Work Ethnographies: Revisiting Breakdown, Relocating Materiality*, ed. Ignaz Strebel, Alain Bovet, and Philippe Sormani (Singapore: Palgrave Macmillan, 2019), 283–312.

33. Houston et al., "Values in Repair," 1407.

34. Houston et al., "Values in Repair," 1408; Rosner and Ames, "Designing for Repair?"

35. Jason Koebler, "Why American Farmers Are Hacking Their Tractors with Ukrainian Firmware," *Motherboard* (blog), March 21, 2017, https://motherboard.vice.com/en _us/article/xykkkd/why-american-farmers-are-hacking-their-tractors-with-ukrainian -firmware; "Stand Up for Your Right to Repair," Repair Association, accessed December 3, 2018, at http://repair.org/stand-up/; "Right to Repair—iFixit," accessed December 3, 2018, at https://ifixit.org/right; Dant, "Inside the Bicycle."

36. Houston and Jackson, "Caring for the 'Next Billion' Mobile Handsets"; Houston et al., "Values in Repair"; Lara Houston, "Unsettled Repair Tools: The 'Death' of the J.A.F. Box" (paper presented at the Maintainers Conference, Stevens Institute of Technology, Hoboken, NJ, 2016), 1–13, https://static1.squarespace.com/ static/56a8e2fca12f446482d67a7a/t/570e9c8f01dbae9c3322fa7d/1460575376524/ Maintainers-Lara-Houston.pdf; Lara Houston, "Mobile Phone Repair Knowledge in Downtown Kampala: Local and Trans-Local Circulations," in *Repair Work Ethnographies Revisiting Breakdown, Relocating Materiality*, ed. Ignaz Strebel, Alain Bovet, and Philippe Sormani (Singapore: Palgrave Macmillan, 2019), 129–160; Steven J. Jackson, Alex Pompe, and Gabriel Krieshok, "Repair Worlds: Maintenance, Repair, and ICT for Development in Rural Namibia," in *Proceedings of the ACM 2012 Conference on Computer Supported Cooperative Work* (New York: ACM Press, 2012), 107; Joshua A. Bell, Briel Kobak, Joel Kuipers, and Amanda Kemble, "Unseen Connections: The Materiality of Cell Phones," *Anthropological Quarterly* 91, no. 2 (2018): 465–484.

37. Houston and Jackson, "Caring for the 'Next Billion' Mobile Handsets," 4.

38. Houston and Jackson, "Caring for the 'Next Billion' Mobile Handsets," 6.

39. Houston and Jackson, "Caring for the 'Next Billion' Mobile Handsets."

40. Houston and Jackson, "Caring for the 'Next Billion' Mobile Handsets," 6–8; Houston, "Unsettled Repair Tools"; Joshua A. Bell, Joel Kuipers, Jacqueline Hazen, Amanda Kemble, and Briel Kobak, "The Materiality of Cell Phone Repair: Re-Making Commodities in Washington, DC," *Anthropological Quarterly* 91, no. 2 (2018): 603–633.

41. Gary Alan Fine, *Tiny Publics: A Theory of Group Action and Culture* (New York: Russell Sage Foundation, 2012); Gary Alan Fine, "Negotiated Orders and Organizational

Cultures," *Annual Review of Sociology* 10 (1984): 239–262; Gary Alan Fine and S. Kleinman, "Network and Meaning: An Interactionist Approach to Structure," *Symbolic Interaction* 6 (1983): 97–110.

Chapter 3

1. Michael Granberry, "20 Years Later, Span Is Loved and Loathed," *Los Angeles Times*, August 2, 1989, http://articles.latimes.com/1989-08-02/local/me-406_1_toll-bridge.

2. Although we are aware *Chicano* is a gendered term and may be considered outdated by some, we use it here when appropriate as part of the name of Chicano Park and to describe related political movements, reflecting the use of the term by many of the activists and artists who are central to the events described in this chapter. When referring to the neighborhood, we use the term *Mexican American* as a more precise description of the community's historic demographics than the broader *Latino/a/x*.

3. This includes actor-network theory (ANT), as developed in STS by Bruno Latour, Michel Callon, and others. The literature on this theory is voluminous, but a useful overview is Bruno Latour, *Reassembling the Social: An Introduction to Actor-Network-Theory* (Oxford: Oxford University Press, 2005). A related STS concept is heterogeneous engineering, as developed by John Law: John Law, "Technology and Heterogeneous Engineering: The Case of Portuguese Expansion," in *The Social Construction of Technological Systems: New Directions in the Sociology and History of Technology*, ed. Wiebe E. Bijker, Thomas P. Hughes, and Trevor J. Pinch (Cambridge, MA: MIT Press, 1987), 111–134, and John Law, *Aircraft Stories: Decentering the Object in Technoscience* (Durham, NC: Duke University Press, 2002). Historian Thomas P. Hughes takes a similar perspective on system builders: Thomas P. Hughes, *Networks of Power: Electrification in Western Society, 1880–1930* (Baltimore: Johns Hopkins University Press, 1983), and Thomas P. Hughes, *Rescuing Prometheus: Four Monumental Projects That Changed the Modern World* (New York: Pantheon Books, 1998). We are particularly influenced by the way elements of ANT are incorporated into an overall historical narrative in the work of sociologist Chandra Mukerji: Chandra Mukerji, *Territorial Ambitions and the Gardens of Versailles* (Cambridge: Cambridge University Press, 1997), and Chandra Mukerji, *Impossible Engineering: Technology and Territoriality on the Canal Du Midi* (Princeton, NJ: Princeton University Press, 2009).

4. Langdon Winner, "Do Artifacts Have Politics?" in *The Social Shaping of Technology*, ed. Donald MacKenzie and Judy Wajcman (Milton Keynes, UK: Open University Press, 1985), 26–38. Although many of Winner's arguments are compatible with the view we take in this book, he makes very strong claims about inherent connections between certain technologies and certain political systems. These claims seem to rely on a view of technological systems as fixed and stable entities once built, a view that our focus on repair and maintenance specifically calls into question.

5. Stephen Graham and Simon Marvin, *Splintering Urbanism: Networked Infrastructures, Technological Mobilities and the Urban Condition* (London: Routledge, 2001). For a detailed case study of how these factors interact, see Andrew R. Highsmith, *Demolition Means Progress: Flint, Michigan, and the Fate of the American Metropolis* (Chicago: University of Chicago Press, 2015).

6. Bruno Latour, *Science in Action: How to Follow Scientists and Engineers through Society* (Cambridge, MA: Harvard University Press, 1987); see also Mukerji, *Territorial Ambitions*, 324.

7. Coronado Historical Association, "A Timeline of Coronado History," 2019, https://coronadohistory.org/calendar/list//education/a-timeline-of-coronado-history/.

8. Unless otherwise noted, the accounts of the building of the San Diego–Coronado Bridge, the history of earthquake engineering practice at Caltrans, the Chicano Park takeover, and the San Diego–Coronado Bridge retrofit project are derived from Benjamin Sims, "On Shifting Ground: Earthquakes, Retrofit and Engineering Culture in California" (PhD diss., University of California, San Diego, 2000), and Benjamin Sims, "Seismic Shifts and Retrofits: Scale and Complexity in the Seismic Retrofit of California Bridges," in *Retrofitting Cities: Priorities, Governance and Experimentation*, ed. Mike Hodson and Simon Marvin (London: Routledge, 2016), 13–33. We cite specific sources only for material that is original to this book.

9. Interview conducted by Sims, April 8, 1997.

10. Interview conducted by Sims, May 29, 1997.

11. Interview conducted by Sims, April 8, 1997.

12. Frank Norris, "Logan Heights: Growth and Change in the Old 'East End,'" *Journal of San Diego History* 29, no. 1 (1983), http://sandiegohistory.org/journal/1983/january/logan/.

13. Kevin Delgado, "A Turning Point: The Conception and Realization of Chicano Park," *Journal of San Diego History* 44, no. 1 (1998), http://www.sandiegohistory.org/journal/1998/january/chicano-3/.

14. Raúl Homero Villa, *Barrio Logos: Space and Place in Urban Chicano Literature and Culture* (Austin: University of Texas Press, 2000), 4. Villa adopts the term from Richard Griswold del Castillo, *The Los Angeles Barrio, 1850–1890: A Social History* (Berkeley: University of California Press, 1979).

15. Villa, *Barrio Logos*, 19–65.

16. Villa, *Barrio Logos*, 66–110; Bob Bolin, Sara Grineski, and Timothy Collins, "The Geography of Despair: Environmental Racism and the Making of South Phoenix, Arizona, USA," *Human Ecology Review* 12, no. 2 (2005): 156–168.

17. Raymond A. Mohl, "Stop the Road: Freeway Revolts in American Cities," *Journal of Urban History* 30, no. 5 (2004): 674–706; Villa, *Barrio Logos*, 66–110.

18. Delgado, "A Turning Point"; Martin D. Rosen and James Fisher, "Chicano Park and the Chicano Park Murals: Barrio Logan, City of San Diego, California," *Public Historian* 23, no. 4 (2001): 93.

19. Juan Felipe Herrera, "Logan Heights and the World," in *100 Great Poems: Classic Poets and Beatnik Freaks* (Words of Wisdom Records, 2012), track 6, quoted in Villa, *Barrio Logos*, 183.

20. Delgado, "A Turning Point"; Mario Barrera, Marilyn Mulford, Juan Felipe Herrera, and Gary Weimberg, *Chicano Park*, VHS tape (Cinema Guild, 1989), https://www.youtube.com/watch?v=hXwZLo8hrp4.

21. Delgado, "A Turning Point"; Villa, *Barrio Logos*, 176–179.

22. Eva Sperling Cockcroft, "The Story of Chicano Park," *Aztlán* 15, no. 1 (1984): 84, quoted in Villa, *Barrio Logos*.

23. Delgado, "A Turning Point."

24. Rosen and Fisher, "Chicano Park and the Chicano Park Murals," 103–105, 106.

25. A map and catalog of the murals, including complete artistic credits, can be found on the website of the Chicano Park Steering Committee: http://chicano-park.com/ChicanoParkMap.jpg.

26. Villa, *Barrio Logos*, 174.

27. Raymond A. Mohl, "The Interstates and the Cities: The U.S. Department of Transportation and the Freeway Revolt, 1966–1973," *Journal of Policy History* 20, no. 2 (2008): 193–226.

28. USGS Earthquake Hazards Program, https://earthquake.usgs.gov/earthquakes/eventpage/nc216859/impact (accessed February 20, 2020).

29. Bill Wallace and Michael Taylor, "State Was Slow to Reinforce the I-880 Pillars, Experts Say," *San Francisco Chronicle*, October 20, 1989.

30. See Sebastián Ureta, "Normalizing Transantiago: On the Challenges (and Limits) of Repairing Infrastructures," *Social Studies of Science* 44, no. 3 (2014): 368–392.

31. Interview with Caltrans District 11 personnel, conducted by Sims, April 1, 1997.

32. Rosen and Fisher, "Chicano Park and the Chicano Park Murals," 107.

33. Interview with Caltrans District 11 personnel, conducted by Sims, April 1, 1997.

34. Rosen and Fisher, "Chicano Park and the Chicano Park Murals."

35. Interview with Caltrans District 11 personnel, conducted by Sims, April 1, 1997; interview with Frieder Seible, conducted by Sims, February 27, 1997.

36. Benjamin Sims, "Concrete Practices: Testing in an Earthquake-Engineering Laboratory," *Social Studies of Science* 29, no. 4 (1999): 483–518; see also Bruno Latour, "Give Me a Laboratory and I Will Raise the World," in *Science Observed*, ed. K. Knorr and M. Mulkay (Beverly Hills: Sage, 1983), 141–170.

37. Interview conducted by Sims, February 27, 1997.

38. Gary Warth, "Chicano Park Named National Historic Landmark," *San Diego Union-Tribune*, January 11, 2017, https://www.sandiegouniontribune.com/news/politics/sd-me-chicano-historic-20170111-story.html.

39. Law, "Technology and Heterogeneous Engineering."

Chapter 4

1. The case study in this introductory section draws on Benjamin Sims and Christopher R. Henke, "Repairing Credibility: Repositioning Nuclear Weapons Knowledge after the Cold War," *Social Studies of Science* 42 (2012): 324–347. See also Stephen I. Schwartz, *Atomic Audit: The Costs and Consequences of U.S. Nuclear Weapons since 1940* (Washington, DC: Brookings Institution Press, 1998), https://www.brookings.edu/book/atomic-audit/; Eric Schlosser, *Command and Control: Nuclear Weapons, the Damascus Accident, and the Illusion of Safety*, reprint ed. (New York: Penguin Books, 2014); Alex Wellerstein, "Maintaining the Bomb," *Restricted Data: The Nuclear Secrecy Blog* (blog), April 8, 2016, http://blog.nuclearsecrecy.com/2016/04/08/maintaining-the-bomb/.

2. Mark Leibovich, "Armageddon Moves Inside the Computer: Los Alamos Is Calculating a New Nuclear Testing Era," *Washington Post*, November 18, 1998.

3. Michael Mann, "The Autonomous Power of the State: Its Origins, Mechanisms and Results," *European Journal of Sociology* 25, no. 2 (1984): 185–213; Michael Mann, *The Sources of Social Power*, vol. 2: *The Rise of Classes and Nation States, 1760–1914* (Cambridge: Cambridge University Press, 1993).

4. William Cronon, *Nature's Metropolis: Chicago and the Great West* (New York: Norton, 1991); Richard White, *The Organic Machine* (New York: Hill and Wang, 1995); Ashley Carse, *Beyond the Big Ditch: Politics, Ecology, and Infrastructure at the Panama Canal* (Cambridge, MA: MIT Press, 2014); Penny Harvey and Hannah Knox, *Roads: An Anthropology of Infrastructure and Expertise* (Ithaca, NY: Cornell University Press, 2015).

5. For example, recent lidar studies have revealed an extensive infrastructure network associated with Mayan civilization and helped establish a close connection between infrastructure system failure and the decline of the Cambodian city of

Angkor. Marcello A. Canuto, Francisco Estrada-Belli, Thomas G. Garrison, Stephen D. Houston, Mary Jane Acuña, Milan Kováč, Damien Marken, "Ancient Lowland Maya Complexity as Revealed by Airborne Laser Scanning of Northern Guatemala," *Science* 361, no. 6409 (2018); Dan Penny, Cameron Zachreson, Roland Fletcher, David Lau, Joseph T. Lizier, Nicholas Fischer, Damian Evans, et al., "The Demise of Angkor: Systemic Vulnerability of Urban Infrastructure to Climatic Variations," *Science Advances* 4, no. 10 (2018).

6. Mann, "The Autonomous Power of the State"; Mann, *The Sources of Social Power*; Patrick Carroll, *Science, Culture, and Modern State Formation* (Berkeley: University of California Press, 2006); Chandra Mukerji, *Modernity Reimagined: An Analytic Guide* (New York: Routledge, 2017).

7. Max Weber, *The Protestant Ethic and the Spirit of Capitalism* (New York: Routledge, 1930); Mukerji, *Modernity Reimagined*.

8. David Harvey, *The Condition of Postmodernity: An Enquiry into the Origins of Cultural Change* (Cambridge, MA: Blackwell, 1990); Paul N. Edwards, "Infrastructure and Modernity: Force, Time, and Social Organization in the History of Sociotechnical Systems," in *Modernity and Technology*, ed. Thomas J. Misa, Philip Brey, and Andrew Feenberg (Cambridge, MA: MIT Press, 2003), 185–225; Brian Larkin, "The Politics and Poetics of Infrastructure," *Annual Review of Anthropology* 42, no. 1 (2013): 327–343; Nikhil Anand, Akhil Gupta, and Hannah Appel, eds., *The Promise of Infrastructure* (Durham, NC: Duke University Press, 2018), Introduction.

9. Mann, "The Autonomous Power of the State," 198.

10. Mann, "The Autonomous Power of the State," 189.

11. Chandra Mukerji, *Territorial Ambitions and the Gardens of Versailles* (Cambridge: Cambridge University Press, 1997); Chandra Mukerji, "Intelligent Uses of Engineering and the Legitimacy of State Power," *Technology and Culture* 44, no. 4 (2003): 655–676; Chandra Mukerji, *Impossible Engineering: Technology and Territoriality on the Canal du Midi* (Princeton, NJ: Princeton University Press, 2009); Chandra Mukerji, "The Landscape Garden as Material Culture: Lessons from France," in *The Oxford Handbook of Material Culture Studies*, ed. Dan Hicks and Mary C. Beaudry (New York: Oxford University Press, 2010), 543–61.

12. Mukerji, *Territorial Ambitions and the Gardens of Versailles*; Mukerji, *Impossible Engineering*; Christopher R. Henke and Thomas F. Gieryn, "Sites of Scientific Practice: The Enduring Importance of Place," in *New Handbook of Science and Technology Studies*, ed. Edward Hackett, Olga Amsterdamska, Michael Lynch, and Judy Wajcman (Cambridge, MA: MIT Press, 2007).

13. Mukerji, *Territorial Ambitions and the Gardens of Versailles*, 39.

14. Mukerji, *Territorial Ambitions and the Gardens of Versailles*; Anne Quito, "The New Fountain at Versailles Was Inspired by 17th-Century Power Games," *Quartz*, 2015. https://qz.com/511706/the-new-fountain-at-versailles-was-inspired-by-17th-century -power-games/. This merging of symbolic and technical resources to demonstrate state power is in some ways quite similar to what has more recently been conveyed through nuclear weapons testing. This convergence was explicit for Homi J. Bhabha, one of the key founders of nuclear research in India (not to be confused with Homi K. Bhabha, the critical theorist). Bhabha carefully planned the grounds of his Atomic Research Establishment, Trombay (now known as the Bhabha Atomic Research Centre), to include an elaborate garden modeled after the gardens at Versailles as part of a larger postcolonial vision for a distinctly Indian appropriation of European research models and high culture. See Stuart Leslie and Indira Chowdhury, "Homi Bhabha, Master Builder of Nuclear India." *Physics Today* 71, no. 9 (2018): 48.

15. Mukerji, *Impossible Engineering*, 136.

16. Mukerji, *Impossible Engineering*, chap. 9.

17. Carroll, *Science, Culture, and Modern State Formation*; Steven Shapin and Simon Schaffer, *Leviathan and the Air-Pump: Hobbes, Boyle, and the Experimental Life* (Princeton, NJ: Princeton University Press, 1985).

18. Carroll, *Science, Culture, and Modern State Formation*, chap. 4.

19. Carroll, *Science, Culture, and Modern State Formation*, 54.

20. Edwards, "Infrastructure and Modernity."

21. Andrew Russell and Lee Vinsel, "Hail the Maintainers," *Aeon*, 2016, https://aeon .co/essays/innovation-is-overvalued-maintenance-often-matters-more; Andrew L. Russell and Lee Vinsel, "After Innovation, Turn to Maintenance," *Technology and Culture* 59, no. 1 (2018): 1–25. See also David Edgerton, *The Shock of the Old: Technology and Global History since 1900* (New York: Oxford University Press, 2007).

22. Emile Durkheim, *The Division of Labor in Society* (New York: Free Press, 1984).

23. Chandra Mukerji, *A Fragile Power: Scientists and the State* (Princeton, NJ: Princeton University Press, 1989).

24. Michel Foucault, *Discipline and Punish: The Birth of the Prison* (New York: Random House, 1979).

25. Max Weber, "Bureaucracy," in *From Max Weber: Essays in Sociology*, ed. H. H. Gerth and C. Wright Mills (New York: Oxford University Press, 1946), 196–244.

26. Thomas Lemke, "New Materialisms: Foucault and the 'Government of Things,'" *Theory, Culture and Society* 32, no. 4 (2015): 3–25.

27. Julie Draskoczy, "The *Put'* of *Perekovka*: Transforming Lives at Stalin's White Sea-Baltic Canal," *Russian Review* 71, no. 1 (2012): 30–31; Cynthia A. Ruder, *Making History for Stalin: The Story of the Belomor Canal* (Gainesville: University Press of Florida, 1998). Thanks to Robert Nemes for suggesting the case of the Belomor canal.

28. Ruder, *Making History for Stalin*, 47.

29. Mukerji, *A Fragile Power*; Thomas F. Gieryn, "Boundary-Work and the Demarcation of Science from Non-Science: Strains and Interests in Professional Ideologies of Science," *American Sociological Review* 48 (1983): 781–795; Thomas F. Gieryn, *Cultural Boundaries of Science: Credibility on the Line* (Chicago: University of Chicago Press, 1999).

30. Mukerji, *A Fragile Power*, 5.

31. Mukerji, *A Fragile Power*, chap. 4. See also Paul N. Edwards, *The Closed World: Computers and the Politics of Discourse in Cold War America* (Cambridge, MA: MIT Press, 1996).

32. Edwards, *The Closed World*.

33. Sims and Henke, "Repairing Credibility," 311.

34. Sims and Henke, "Repairing Credibility," 340.

35. Larkin, "The Politics and Poetics of Infrastructure," 333.

36. Benedict Anderson, *Imagined Communities: Reflections on the Origin and Spread of Nationalism* (London: Verso, 1983), 8.

37. Anderson, *Imagined Communities*, chap. 10.

38. Anderson, *Imagined Communities*, 38.

39. Anderson, *Imagined Communities*, chap. 4.

40. Joseph Masco, *The Nuclear Borderlands: The Manhattan Project in Post–Cold War New Mexico* (Princeton, NJ: Princeton University Press, 2006), 3, emphasis in original.

41. Michael Billig, *Banal Nationalism* (Thousand Oaks, CA: Sage, 1995).

42. Billig, *Banal Nationalism*, 95; Rhys Jones and Peter Merriman, "Hot, Banal and Everyday Nationalism: Bilingual Road Signs in Wales," *Political Geography* 28, no. 3 (2009): 164–173; Larkin, "The Politics and Poetics of Infrastructure"; Peter Merriman and Rhys Jones, "Nations, Materialities and Affects," *Progress in Human Geography* 41, no. 5 (2017): 600–617.

43. Masco, *The Nuclear Borderlands*; Kelly Moore, *Disrupting Science: Social Movements, American Scientists, and the Politics of the Military, 1945–1975* (Princeton, NJ: Princeton University Press, 2008).

44. James C. Scott, *Seeing Like a State: How Certain Schemes to Improve the Human Condition Have Failed* (New Haven: Yale University Press, 1998).

45. Scott, *Seeing Like a State*, chap. 4.

46. Scott, *Seeing Like a State*, 125.

47. James C. Scott, *The Moral Economy of the Peasant: Rebellion and Subsistence in Southeast Asia* (New Haven: Yale University Press, 1976); James C. Scott, *Weapons of the Weak: Everyday Forms of Peasant Resistance* (New Haven: Yale University Press, 1987).

48. Larissa Pires, "Gender in the Modernist City: Shaping Power Relations and National Identity with the Construction of Brasilia" (PhD diss., Iowa State University, 2013), chap. 6.

49. Pires, "Gender in the Modernist City," 305. See also James Holston, *The Modernist City: An Anthropological Critique of Brasilia* (Chicago: University of Chicago Press, 1989); Joseph Rykwert, *The Seduction of Place: The History and Future of the City* (New York: Oxford University Press, 2004).

50. Antina von Schnitzler, "Infrastructure, Apartheid Technopolitics, and Temporalities of 'Transition,'" in *The Promise of Infrastructure*, ed. Nikhil Anand, Akhil Gupta, and Hannah Appel (Durham, NC: Duke University Press, 2018), 135; Antina von Schnitzler, *Democracy's Infrastructure: Techno-Politics and Protest after Apartheid*, reprint ed. (Princeton, NJ: Princeton University Press, 2016).

51. Mann, "The Autonomous Power of the State."

52. Harvey, *The Condition of Postmodernity*; Bruno Latour, *We Have Never Been Modern* (Cambridge, MA: Harvard University Press, 1993).

53. David Harvey, *A Brief History of Neoliberalism* (Oxford: Oxford University Press, 2007); Dominic Boyer, "Infrastructure, Potential Energy, Revolution," in *The Promise of Infrastructure*, ed. Nikhil Anand, Akhil Gupta, and Hannah Appel (Durham, NC: Duke University Press, 2018), 223–243.

54. Harvey, *A Brief History of Neoliberalism*.

55. von Schnitzler, *Democracy's Infrastructure*.

56. Elana Shever, *Resources for Reform: Oil and Neoliberalism in Argentina* (Palo Alto, CA: Stanford University Press, 2012).

57. Timothy Mitchell, *Carbon Democracy: Political Power in the Age of Oil* (London: Verso Books, 2011); Boyer, "Infrastructure, Potential Energy, Revolution."

58. Boyer, "Infrastructure, Potential Energy, Revolution," 223; Steve Graham and Simon Marvin, *Splintering Urbanism: Networked Infrastructures, Technological Mobilities and the Urban Condition* (London; New York: Routledge, 2001), chap. 3.

59. Steven J. Jackson, "Rethinking Repair," in *Media Technologies: Essays on Communication, Materiality and Society*, ed. Tarleton Gillespie, Pablo Boczkowski, and Kirsten Foot (Cambridge, MA: MIT Press, 2014), 221–239.

Chapter 5

1. Sarah K. Lowder, Jakob Skoet, and Saumya Singh, "What Do We Really Know about the Number and Distribution of Farms and Family Farms Worldwide? Background Paper for *The State of Food and Agriculture 2014*" (Rome: United Nations, Food and Agriculture Organization, 2014). In China and India, hundreds of millions of rural farmers cultivate small plots of land averaging two hectares (five acres) or less; however, Lowder, Skoet, and Singh also present some evidence that farm consolidation is a recent trend in Asia, after a period (roughly 1960–1980) when farm sizes decreased in most of the developing world (page 10).

2. D. M. Spielmaker, "Growing a Nation: The Story of American Agriculture," 2018, https://www.agclassroom.org/gan/timeline/index.htm; National Agricultural Statistics Service, "USDA Census of Agriculture Historical Archive," 2018, http://agcensus .mannlib.cornell.edu/AgCensus/homepage.do.

3. William Cronon, *Nature's Metropolis: Chicago and the Great West* (New York: Norton, 1991); Lawrence Busch, *Standards: Recipes for Reality* (Cambridge, MA: MIT Press, 2011); Warren J. Belasco, *Appetite for Change: How the Counterculture Took on the Food Industry*, 2nd ed. (Ithaca, NY: Cornell University Press, 2007); Marion Nestle, *Food Politics: How the Food Industry Influences Nutrition and Health* (Berkeley: University of California Press, 2002).

4. Charles E. Rosenberg, "Rationalization and Reality in the Shaping of American Agricultural Research, 1875–1914," *Social Studies of Science* 7 (1977): 401–422; David B. Danbom, *The Resisted Revolution: Urban America and the Industrialization of Agriculture, 1900–1930* (Ames: Iowa State University Press, 1979); David B. Danbom, *Born in the Country: A History of Rural America* (Baltimore, MD: Johns Hopkins University Press, 1995); Alan I. Marcus, *Agricultural Science and the Quest for Legitimacy: Farmers, Agricultural Colleges, and Experiment Stations, 1870–1890* (Ames: Iowa State University Press, 1985); Lawrence Busch and William B. Lacy, *Science, Agriculture, and the Politics of Research* (Boulder, CO: Westview Press, 1983); Willard W. Cochrane, *The Development of American Agriculture: A Historical Analysis*, 2nd ed. (Minneapolis: University of Minnesota Press, 1993); Christopher R. Henke, *Cultivating Science, Harvesting Power: Science and Industry in California Agriculture* (Cambridge, MA: MIT Press, 2008).

5. Cynthia Nickerson and Allison Borchers, "How Is Land in the United States Used? A Focus on Agricultural Land," 2012, https://www.ers.usda.gov/amber-waves/ 2012/march/data-feature-how-is-land-used/; USDA ERS, "Ag and Food Sectors and the Economy," 2018, https://www.ers.usda.gov/data-products/ag-and-food-statistics

-charting-the-essentials/ag-and-food-sectors-and-the-economy.aspx; Dave Merrill and Lauren Leatherby, "Here's How America Uses Its Land," *Bloomberg*, July 31, 2018, https://www.bloomberg.com/graphics/2018-us-land-use/.

6. Alisha Coleman-Jensen, Matthew P. Rabbitt, Christian A. Gregory, and Anita Singh, "Household Food Security in the United States in 2016" (Washington, DC: USDA ERS, 2017), v–vi; Marion Nestle, *Food Politics: How the Food Industry Influences Nutrition and Health* (Berkeley: University of California Press, 2002); Michael Pollan, *The Omnivore's Dilemma: A Natural History of Four Meals* (New York: Penguin, 2006); Alison Hope Alkon and Julian Agyeman, *Cultivating Food Justice: Race, Class, and Sustainability* (Cambridge, MA: MIT Press, 2011); Kathryn Edin, Melody Boyd, James Mabli, Jim Ohls, Julie Worthington, Sara Greene, et al., "SNAP Food Security In-Depth Interview Study" (Washington, DC: US Department of Agriculture, Food and Nutrition Service, 2013); Kathryn J. Edin and H. Luke Shaefer, *$2.00 a Day: Living on Almost Nothing in America* (New York: Houghton Mifflin Harcourt, 2015).

7. US EPA, "National Summary of State Information: Water Quality Assessment and TMDL Information," 2018, https://ofmpub.epa.gov/waters10/attains_nation_cy .control#prob_source.

8. J. Reilly, F. Tubiello, B. McCarl, D. Abler, R. Darwin, K. Fuglie, S. Hollinger, et al., "U.S. Agriculture and Climate Change: New Results," *Climatic Change* 57, no. 1/2 (2003): 43–67; C. L. Walthall, J. Hatfield, P. Backlund, and L. Lengnick, "Climate Change and Agriculture in the United States: Effects and Adaptation" (Washington, DC: USDA, 2012), https://www.usda.gov/oce/climate_change/effects_2012/CC%20 and%20Agriculture%20Report%20(02-04-2013)b.pdf; Christopher B. Field, Vicente R. Barros, and Intergovernmental Panel on Climate Change, eds., *Climate Change 2014: Impacts, Adaptation, and Vulnerability: Working Group II Contribution to the Fifth Assessment Report of the Intergovernmental Panel on Climate Change* (Cambridge: Cambridge University Press, 2014); Laura Lengnick, *Resilient Agriculture: Cultivating Food Systems for a Changing Climate* (Philadelphia: New Society Publishers, 2015); US EPA, "Sources of Greenhouse Gas Emissions," Overviews and Factsheets, US EPA, December 29, 2015, https://www.epa.gov/ghgemissions/sources-greenhouse-gas-emissions; US Global Change Research Program, "Fourth National Climate Assessment" (Washington, DC, 2018), https://nca2018.globalchange.gov/chapter/10.

9. Henke, *Cultivating Science, Harvesting Power*, chap. 6.

10. Paul N. Edwards, *A Vast Machine: Computer Models, Climate Data, and the Politics of Global Warming* (Cambridge, MA: MIT Press, 2010).

11. For a detailed discussion of the uses and misuses of reflexivity in sociology and STS, see Michael Lynch, "Against Reflexivity as an Academic Virtue and Source of Privileged Knowledge," *Theory, Culture and Society* 17, no. 3 (2000): 26–54.

12. Steven J. Jackson, "Rethinking Repair," in *Media Technologies: Essays on Communication, Materiality and Society*, ed. Tarleton Gillespie, Pablo Boczkowski, and Kirsten Foot (Cambridge, MA: MIT Press, 2014), 221–239.

13. Ulrich Beck, *Risk Society: Towards a New Modernity* (London: Sage, 1992).

14. See also Brian Wynne, "Sheepfarming after Chernobyl: A Case Study in Communicating Scientific Information," *Environment* 31 (1989): 10–15, 33–39; Brian Wynne, "May the Sheep Safely Graze? A Reflexive View of the Expert-Lay Divide," in *Risk, Environment, and Modernity: Towards a New Ecology*, ed. Scott Lash, Bronislaw Szerszynski, and Brian Wynne (Thousand Oaks, CA: Sage, 1996), 44–83; Olga Kuchinskaya, *The Politics of Invisibility: Public Knowledge about Radiation Health Effects after Chernobyl* (Cambridge, MA: MIT Press, 2014).

15. Beck, *Risk Society*; Anthony Giddens, "Risk and Responsibility," *Modern Law Review* 62, no. 1 (1999): 1–10; Anthony Giddens, *Runaway World: How Globalization Is Reshaping Our Lives* (New York: Routledge, 2000); Ulrich Beck, Anthony Giddens, and Scott Lash, *Reflexive Modernization: Politics, Tradition and Aesthetics in the Modern Social Order* (Cambridge: Polity Press, 1994); Ulrich Beck, Wolfgang Bonss, and Christoph Lau, "The Theory of Reflexive Modernization: Problematic, Hypotheses and Research Programme," *Theory, Culture and Society* 20, no. 2 (2003): 1–33.

16. Bruno Latour, *We Have Never Been Modern* (Cambridge, MA: Harvard University Press, 1993).

17. Paul N. Edwards, "Meteorology as Infrastructural Globalism," *Osiris* 21, no. 1 (2006): 230.

18. See also Paul N. Edwards, *A Vast Machine: Computer Models, Climate Data, and the Politics of Global Warming* (Cambridge, MA: MIT Press, 2010).

19. Benjamin Sims, "Making Technological Timelines: Anticipatory Repair and Testing in High Performance Scientific Computing," *Continent* 6, no. 1 (2017): 81–84; see also Stephanie B. Steinhardt and Steven J. Jackson, "Anticipation Work: Cultivating Vision in Collective Practice," in *Proceedings of the 18th ACM Conference on Computer Supported Cooperative Work and Social Computing* (New York: ACM, 2015), 443–453.

20. World Commission on Environment and Development, *Our Common Future* (New York: Oxford University Press, 1987).

21. Robert W. Kates, Thomas M. Parris, and Anthony A. Leiserowitz, "What Is Sustainable Development? Goals, Indicators, Values, and Practice," *Environment* 47, no. 3 (2005): 9–10.

22. Andrew Dobson, ed., *Fairness and Futurity: Essays on Environmental Sustainability and Social Justice* (New York: Oxford University Press, 1999); Kates, Parris, and Leiserowitz, "What Is Sustainable Development?"; Magnus Boström, "A Missing Pillar?

Challenges in Theorizing and Practicing Social Sustainability," *Sustainability: Science, Practice, and Policy* 8, no. 1 (2012): 3–14.

23. Tom Athanasiou, "The Age of Greenwashing," *Capitalism, Nature, Socialism* 7 (1996): 1–36.

24. Kates et al., "What Is Sustainable Development?," 20.

25. Association for the Advancement of Sustainability in Higher Education, "AASHE STARS," 2016, https://stars.aashe.org/; Second Nature, "The Climate Leadership Commitments," Second Nature, 2016, http://secondnature.org/what-we-do/climate-leadership/.

26. This example is excerpted from Christopher R. Henke, "The Sustainable University: Repair as Maintenance and Transformation," *Continent* 6, no. 1 (2017): 40–45.

27. U.S. Green Building Council, "LEED | U.S. Green Building Council," 2016, http://www.usgbc.org/leed.

28. Natalie Sportelli, "Colgate University's Trudy Fitness Center Awarded Gold LEED Certification," *Colgate University News* (blog), June 27, 2012, http://news.colgate.edu/2012/06/trudy-fitness-center-awarded-gold-leed-certification.html/; Colgate University, "Colgate University Green Building Standards," 2014, https://docs.google.com/viewer?url=http%3A%2F%2Fwww.colgate.edu%2Fdocs%2Fdefault-source%2Fdefault-document-library%2FRead-More-365849.pdf%3Fsfvrsn%3D3.

29. Andrew Ross, "Universities and the Urban Growth Machine," *Dissent Magazine*, October 4, 2012, http://www.dissentmagazine.org/online_articles/universities-and-the-urban-growth-machine.

30. Association for the Advancement of Sustainability in Higher Education, "AASHE STARS," 2016, https://stars.aashe.org/.

31. Geoff Bowker and Susan Leigh Star, *Sorting Things Out: Classification and Practice* (Cambridge, MA: MIT Press, 1999); Martha Lampland and Susan Leigh Star, eds., *Standards and Their Stories: How Quantifying, Classifying and Formalizing Practices Shape Everyday Life* (Ithaca, NY: Cornell University Press, 2008); Busch, *Standards*.

32. Second Nature's Resilience Commitment is one model that asks universities to consider such partnerships in order to facilitate "climate adaptation and community capacity-building to deal with a changing climate and resulting extremes." Second Nature. "The Presidents' Climate Leadership Commitments," 2019, https://secondnature.org/signatory-handbook/the-commitments/.

33. As this book goes to press, Colgate is in fact considering local investments in reforestation as part of a package of climate offsets to mitigate the university's climate footprint, especially the impacts that are harder to reduce through reduction or replacement efforts. Colgate University, "Carbon Neutrality and Carbon Offsets FAQ," 2019,

https://200.colgate.edu/looking-forward/our-sustainability-commitment/carbon -neutrality-and-carbon-offsets-faq.

34. Philip Sirianni and Michael O'Hara, "Do Actions Speak as Loud as Words? Commitments to 'Going Green' on Campus," *Contemporary Economic Policy* 32, no. 2 (2014): 503–519; Michael O'Hara and Philip Sirianni, "Carbon Efficiency of US Colleges and Universities: A Nonparametric Assessment," *Applied Economics* 49, no. 11 (2017): 1083–1097; Chandra Russo and Andrew Pattison, "The Pitfalls and Promises of Climate Action Plans: Transformative Resilience Strategy in U.S. Cities," in *Resilience, Environmental Justice and the City*, ed. Beth Schaefer Caniglia, Manuel Vallée, and Beatrice Frank (New York: Routledge, 2017).

35. Jonathan Bach, "China's Infrastructural Fix," *Limn*, no. 7 (2016), https://limn.it/articles/chinas-infrastructural-fix/.

36. Federico Demaria, "The Rise—and Future—of the Degrowth Movement," *Ecologist*, 2018, https://theecologist.org/2018/mar/27/rise-and-future-degrowth-movement; Giorgos Kallis, *Degrowth* (Newcastle upon Tyne: Agenda Publishing, 2018).

37. Kallis, *Degrowth*.

38. Paul Kingsworth and Dougald Hine, "The Manifesto," Dark Mountain, 2009, https://dark-mountain.net/about/manifesto. Thanks to Lee Vinsel for drawing our attention to the existence of this group.

39. Kingsworth and Hine, "The Manifesto"; Hana Librová and Vojtěch Pelikán, "Ethical Motivations and the Phenomenon of Disappointment in Two Types of Environmental Movements: Neo-Environmentalism and the Dark Mountain Project," *Environmental Values* 25, no. 2 (2016): 167–193.

40. Andrew L. Russell and Lee Vinsel, "After Innovation, Turn to Maintenance," *Technology and Culture* 59, no. 1 (2018): 13.

41. Steven J. Jackson, "Repair as Transition: Time, Materiality, and Hope," in *Repair Work Ethnographies: Revisiting Breakdown, Relocating Materiality*, ed. Ignaz Strebel, Alain Bovet, and Philippe Sormani (Singapore: Palgrave Macmillan, 2019), 346.

42. Lynch, "Against Reflexivity."

43. IPCC, *Climate Change 2014: Mitigation of Climate Change. Contribution of Working Group III to the Fifth Assessment Report of the Intergovernmental Panel on Climate Change* (Cambridge: Cambridge University Press, 2014).

44. Lorna A. Greening, David L. Greene, and Carmen Difiglio, "Energy Efficiency and Consumption—the Rebound Effect—a Survey," *Energy Policy* 28, no. 6 (2000): 389–401.

Bibliography

Abbate, Janet. *Inventing the Internet:* Cambridge, MA: MIT Press, 1999.

Alkon, Alison Hope, and Julian Agyeman. *Cultivating Food Justice: Race, Class, and Sustainability.* Cambridge, MA: MIT Press, 2011.

Anand, Nikhil, Akhil Gupta, and Hannah Appel, eds. *The Promise of Infrastructure.* Durham, NC: Duke University Press, 2018.

Anderson, Benedict. *Imagined Communities: Reflections on the Origin and Spread of Nationalism.* London: Verso, 1983.

Ashford, Blake E., and Glen E. Kreiner. "'How Can You Do It?' Dirty Work and the Challenge of Constructing a Positive Identity." *Academy of Management Review* 24, no. 3 (1999): 413–434.

Association for the Advancement of Sustainability in Higher Education. "AASHE STARS." 2016. https://stars.aashe.org/.

Athanasiou, Tom. "The Age of Greenwashing." *Capitalism, Nature, Socialism* 7 (1996): 1–36.

Bach, Jonathan. "China's Infrastructural Fix." *Limn,* no. 7 (2016). https://limn.it/articles/chinas-infrastructural-fix/.

Barad, Karen. "Posthumanist Performativity: Toward an Understanding of How Matter Comes to Matter." *Signs: Journal of Women in Culture and Society* 28, no. 3 (2003): 801–831.

Barrera, Mario, Marilyn Mulford, Juan Felipe Herrera, and Gary Weimberg. *Chicano Park.* VHS tape. Cinema Guild, 1989. https://www.youtube.com/watch?v=hXwZLo8hrp4.

Barry, John M. *Rising Tide: The Great Mississippi Flood of 1927 and How It Changed America.* New York: Simon & Schuster, 1998.

Batt, William H. "Infrastructure: Etymology and Import." *Journal of Professional Issues in Engineering* 110, no. 1 (1984): 1–6.

Beck, Ulrich. *Risk Society: Towards a New Modernity.* London: Sage, 1992.

Beck, Ulrich, Wolfgang Bonss, and Christoph Lau. "The Theory of Reflexive Modernization: Problematic, Hypotheses and Research Programme." *Theory, Culture and Society* 20, no. 2 (2003): 1–33.

Beck, Ulrich, Anthony Giddens, and Scott Lash. *Reflexive Modernization: Politics, Tradition and Aesthetics in the Modern Social Order.* Cambridge, UK: Polity Press, 1994.

Belasco, Warren J. *Appetite for Change: How the Counterculture Took on the Food Industry.* 2nd ed. Ithaca, NY: Cornell University Press, 2007.

Bell, Joshua A., Briel Kobak, Joel Kuipers, and Amanda Kemble. "Unseen Connections: The Materiality of Cell Phones." *Anthropological Quarterly* 91, no. 2 (2018): 465–484.

Bell, Joshua A., Joel Kuipers, Jacqueline Hazen, Amanda Kemble, and Briel Kobak. "The Materiality of Cell Phone Repair: Re-Making Commodities in Washington, DC." *Anthropological Quarterly* 91, no. 2 (2018): 603–633.

Bellacasa, María Puig de la. "Matters of Care in Technoscience: Assembling Neglected Things." *Social Studies of Science* 41, no. 1 (2011): 85–106.

Bellacasa, María Puig de la. "'Nothing Comes without Its World': Thinking with Care." *Sociological Review* 60, no. 2 (2012): 197–216.

Belluck, Pam. "Chilly at Work? Office Formula Was Devised for Men." *New York Times*, August 3, 2015. https://www.nytimes.com/2015/08/04/science/chilly-at-work-a-decades-old-formula-may-be-to-blame.html.

Berger, Peter L., and Thomas Luckmann. *The Social Construction of Reality: A Treatise in the Sociology of Knowledge.* Harmondsworth: Penguin, 1971.

Bialik, Carl. "We Still Don't Know How Many People Died Because of Katrina." *FiveThirtyEight* (blog), August 26, 2015. https://fivethirtyeight.com/features/we-still-dont-know-how-many-people-died-because-of-katrina/.

Bijker, Wiebe E., Thomas P. Hughes, and Trevor J. Pinch, eds. *The Social Construction of Technological Systems: New Directions in the Sociology and History of Technology.* Cambridge, MA: MIT Press, 1987.

Bijker, Wiebe E., and John Law, eds. *Shaping Technology/Building Society: Studies in Sociotechnical Change.* Cambridge, MA: MIT Press, 1992.

Billig, Michael. *Banal Nationalism.* Thousand Oaks, CA: Sage, 1995.

Bolin, Bob, Sara Grineski, and Timothy Collins. "The Geography of Despair: Environmental Racism and the Making of South Phoenix, Arizona, USA." *Human Ecology Review* 12, no. 2 (2005): 156–168.

Boström, Magnus. "A Missing Pillar? Challenges in Theorizing and Practicing Social Sustainability." *Sustainability: Science, Practice, and Policy* 8, no. 1 (2012): 3–14.

Bowker, Geoffrey C. *Science on the Run: Information Management and Industrial Geophysics at Schlumberger, 1920–1940.* Cambridge, MA: MIT Press, 1994.

Bowker, Geoffrey C., and Susan Leigh Star. *Sorting Things Out: Classification and Its Consequences.* Cambridge, MA: MIT Press, 1999.

Boyer, Dominic. "Infrastructure, Potential Energy, Revolution." In *The Promise of Infrastructure*, edited by Nikhil Anand, Akhil Gupta, and Hannah Appel, 223–243. Durham, NC: Duke University Press, 2018.

Brunkard, Joan, Gonza Namulanda, and Raoult Ratard. "Hurricane Katrina Deaths, Louisiana, 2005." *Disaster Medicine and Public Health Preparedness* 2, no. 4 (2008): 215–223.

Busch, Lawrence. *Standards: Recipes for Reality.* Cambridge, MA: MIT Press, 2011.

Busch, Lawrence, and William B. Lacy. *Science, Agriculture, and the Politics of Research.* Boulder, CO: Westview Press, 1983.

Callon, Michel. "Some Elements of a Sociology of Translation: Domestication of the Scallops and the Fishermen of St. Brieuc Bay." In *Power, Action, and Belief: A New Sociology of Knowledge?* edited by John Law, 196–233. London: Routledge, Kegan, and Paul, 1986.

Canuto, Marcello A., Francisco Estrada-Belli, Thomas G. Garrison, Stephen D. Houston, Mary Jane Acuña, Milan Kováč, et al. "Ancient Lowland Maya Complexity as Revealed by Airborne Laser Scanning of Northern Guatemala." *Science* 361, no. 6409 (2018): eaau0137.

Carroll, Patrick. *Science, Culture, and Modern State Formation.* Berkeley: University of California Press, 2006.

Carse, Ashley. *Beyond the Big Ditch: Politics, Ecology, and Infrastructure at the Panama Canal.* Cambridge, MA: MIT Press, 2014.

Carse, Ashley. "Keyword: Infrastructure: How a Humble French Engineering Term Shaped the Modern World." In *Infrastructures and Social Complexity: A Companion*, edited by Penelope Harvey, Casper Bruun Jensen, and Atsuro Morita, 27–39. London: Routledge, 2016.

Catton, William R., Jr., and Riley E. Dunlap. "Environmental Sociology: A New Paradigm." *American Sociologist* 13, no. 1 (1978): 41–49.

Chalabi, Mona. "Dear Mona, How Many Flight Attendants Are Men?" *FiveThirtyEight* (blog), October 3, 2014. https://fivethirtyeight.com/datalab/dear-mona-how-many-flight-attendants-are-men/.

Cochrane, Willard W. *The Development of American Agriculture: A Historical Analysis.* 2nd ed. Minneapolis: University of Minnesota Press, 1993.

Cockcroft, Eva Sperling. "The Story of Chicano Park." *Aztlán* 15, no. 1 (1984): 79–103.

Cohn, Marisa. "Convivial Decay: Entangled Lifetimes in a Geriatric Infrastructure." In *Proceedings of the 19th ACM Conference on Computer-Supported Cooperative Work and Social Computing*, 1511–1523. New York: ACM, 2016.

Cohn, Marisa. "'Lifetime Issues': Temporal Relations of Design and Maintenance." *Continent* 6, no. 1 (2017): 4–12.

Coleman-Jensen, Alisha, Matthew P. Rabbitt, Christian A. Gregory, and Anita Singh. "Household Food Security in the United States in 2016." Washington, DC: USDA ERS, 2017.

Colgate University. "Colgate University Green Building Standards." 2014. https://docs.google.com/viewer?url=http%3A%2F%2Fwww.colgate.edu%2Fdocs%2Fdefault-source%2Fdefault-document-library%2FRead-More-365849.pdf%3Fsfvrsn%3D3.

Collier, Stephen J., and Andrew Lakoff. "The Vulnerability of Vital Systems: How 'Critical Infrastructure' Became a Security Problem." In *The Politics of Securing the Homeland: Critical Infrastructure, Risk and Securitisation*, edited by Myriam Dunn and Kristian Soby Kristensen. London: Routledge, 2008.

Coronado Historical Association. "A Timeline of Coronado History." 2019. https://coronadohistory.org/calendar/list//education/a-timeline-of-coronado-history/.

Cronon, William. *Nature's Metropolis: Chicago and the Great West.* New York: Norton, 1991.

Crutzen, Paul J. "Geology of Mankind." *Nature* 415 (2002): 23.

Danbom, David B. *Born in the Country: A History of Rural America.* Baltimore, MD: Johns Hopkins University Press, 1995.

Danbom, David B. *The Resisted Revolution: Urban America and the Industrialization of Agriculture, 1900–1930.* Ames: Iowa State University Press, 1979.

Dant, Tim. "Inside the Bicycle: Repair Knowledge for All." In *Repair Work Ethnographies: Revisiting Breakdown, Relocating Materiality*, edited by Ignaz Strebel, Alain Bovet, and Philippe Sormani, 283–312. Singapore: Palgrave Macmillan, 2019.

Dant, Tim. *Materiality and Society.* New York: Open University Press, 2005.

Dant, Tim. "The Work of Repair: Gesture, Emotion, and Sensual Knowledge." *Sociological Research Online* 15, no. 3 (2010): 97–118.

Delgado, Kevin. "A Turning Point: The Conception and Realization of Chicano Park." *Journal of San Diego History* 44, no. 1 (1998). http://www.sandiegohistory.org/journal/1998/january/chicano-3/.

Demaria, Federico. "The Rise—and Future—of the Degrowth Movement." *Ecologist*, March 27, 2018. http://theecologist.org/2018/mar/27/rise-and-future-degrowth-movement.

Denis, Jérôme, and David Pontille. "Maintenance Work and the Performativity of Urban Inscriptions: The Case of Paris Subway Signs." *Environment and Planning D: Society and Space* 32, no. 3 (2014): 404–416.

Denis, Jérôme, and David Pontille. "Material Ordering and the Care of Things." *Science, Technology, and Human Values* 40, no. 3 (2015): 338–367.

Dibble, Sandra, and Kristina Davis. "Officer Punched, Tensions Flare at 'Patriot Picnic' at Chicano Park." *San Diego Union-Tribune*, February 3, 2018. https://www.sandiegouniontribune.com/news/politics/sd-me-patriot-picnic-20180131-story.html.

Dobson, Andrew, ed. *Fairness and Futurity: Essays on Environmental Sustainability and Social Justice*. New York: Oxford University Press, 1999.

Douglas, Mary. *Purity and Danger: An Analysis of Concepts of Pollution and Taboo*. London: Routledge & Kegan Paul, 1966.

Draskoczy, Julie. "The *Put'* of *Perekovka*: Transforming Lives at Stalin's White Sea-Baltic Canal." *Russian Review* 71, no. 1 (2012): 30–48.

Duffy, Mignon. "Doing the Dirty Work: Gender, Race, and Reproductive Labor in Historical Perspective." *Gender and Society* 21, no. 3 (2007): 313–336.

Dunlap, Riley E., and William R. Catton. "Struggling with Human Exemptionalism: The Rise, Decline, and Revitalization of Environmental Sociology." *American Sociologist* 25, no. 1 (1994): 5–30.

Durkheim, Emile. *The Division of Labor in Society*. Translated by W. D. Halls. New York: Free Press, 1984.

Edgerton, David. *The Shock of the Old: Technology and Global History Since 1900*. New York: Oxford University Press, 2007.

Edin, Kathryn, Melody Boyd, James Mabli, Jim Ohls, Julie Worthington, Sara Greene, Nicholas Redel, et al. "SNAP Food Security In-Depth Interview Study." Washington, DC: US Department of Agriculture, Food and Nutrition Service, 2013.

Edin, Kathryn J., and H. Luke Shaefer. *$2.00 a Day: Living on Almost Nothing in America*. New York: Houghton Mifflin Harcourt, 2015.

Edwards, Paul N. *The Closed World: Computers and the Politics of Discourse in Cold War America*. Cambridge, MA: MIT Press, 1996.

Edwards, Paul N. "Infrastructure and Modernity: Force, Time, and Social Organization in the History of Sociotechnical Systems." In *Modernity and Technology*, edited by Thomas J. Misa, Philip Brey, and Andrew Feenberg, 185–225. Cambridge, MA: MIT Press, 2003.

Edwards, Paul N. "Meteorology as Infrastructural Globalism." *Osiris* 21, no. 1 (2006): 229–250.

Edwards, Paul N. *A Vast Machine: Computer Models, Climate Data, and the Politics of Global Warming.* Cambridge, MA: MIT Press, 2010.

Edwards, Paul N., Geoffrey Bowker, Steven Jackson, and Robin Williams. "Introduction: An Agenda for Infrastructure Studies." *Journal of the Association for Information Systems* 10, no. 5 (2009): 364–374.

Edwards, Paul N., Steven J. Jackson, Geoffrey C. Bowker, and Cory P. Knobel. "Understanding Infrastructure: Dynamics, Tensions, and Design." Washington, DC: National Science Foundation, January 2007.

Ellis, Erle C., Dorian Q. Fuller, Jed O. Kaplan, and Wayne G. Lutters. "Dating the Anthropocene: Towards an Empirical Global History of Human Transformation of the Terrestrial Biosphere." *Elementa: Science of the Anthropocene* 1 (December 4, 2013): 000018.

Ellis, Erle C., and Navin Ramankutty. "Putting People in the Map: Anthropogenic Biomes of the World." *Frontiers in Ecology and the Environment* 6, no. 8 (2008): 439–447.

Fanger, P. O. "Assessment of Man's Thermal Comfort in Practice." *British Journal of Industrial Medicine* 30 (1973): 313–324.

Fanger, P. O. *Thermal Comfort: Analysis and Applications in Environmental Engineering.* Copenhagen: Danish Technical Press, 1970.

Field, Christopher B., Vicente R. Barros, and Intergovernmental Panel on Climate Change, eds. *Climate Change 2014: Impacts, Adaptation, and Vulnerability: Working Group II Contribution to the Fifth Assessment Report of the Intergovernmental Panel on Climate Change.* Cambridge: Cambridge University Press, 2014.

Fine, Gary Alan. "Negotiated Orders and Organizational Cultures." *Annual Review of Sociology* 10 (1984): 239–262.

Fine, Gary Alan. *Tiny Publics: A Theory of Group Action and Culture.* New York: Russell Sage Foundation, 2012.

Fine, Gary Alan, and S. Kleinman. "Network and Meaning: An Interactionist Approach to Structure." *Symbolic Interaction* 6 (1983): 97–110.

Finney, Stanley C., and Asier Hilario. "GSSPs as International Geostandards and as Global Geoheritage." In *Geoheritage*, edited by Emmanuel Reynard and José Brilha, 179–189. Amsterdam: Elsevier, 2018.

Fiore-Gartland, Brittany. "Technological Residues." *Continent* 6, no. 1 (2017): 25–29.

Fitzsimmons, Emma G. "What Trump, Clinton and Voters Agreed On: Better Infrastructure." *New York Times*. November 9, 2016.

Foucault, Michel. *Discipline and Punish: The Birth of the Prison*. New York: Random House, 1979.

Freudenburg, William R., Robert B. Gramling, Shirley Laska, and Kai Erikson. *Catastrophe in the Making: The Engineering of Katrina and the Disasters of Tomorrow*. Washington, DC: Island Press, 2009.

Furlong, Kathryn. "STS beyond the 'Modern Infrastructure Ideal': Extending Theory by Engaging with Infrastructure Challenges in the South." *Technology in Society* 38 (2014): 139–147.

Fussell, Elizabeth, Narayan Sastry, and Mark VanLandingham. "Race, Socioeconomic Status, and Return Migration to New Orleans after Hurricane Katrina." *Population and Environment* 31, no. 1–3 (2010): 20–42.

Giddens, Anthony. "Risk and Responsibility." *Modern Law Review* 62, no. 1 (1999): 1–10.

Giddens, Anthony. *Runaway World: How Globalization Is Reshaping Our Lives*. New York: Routledge, 2000.

Gieryn, Thomas F. "Boundary-Work and the Demarcation of Science from Non-Science: Strains and Interests in Professional Ideologies of Science." *American Sociological Review* 48 (1983): 781–795.

Gieryn, Thomas F. *Cultural Boundaries of Science: Credibility on the Line*. Chicago: University of Chicago Press, 1999.

Glenn, Evelyn Nakano. "Creating a Caring Society." *Contemporary Sociology* 29, no. 1 (2000): 84–94.

Goffman, Erving. *Interaction Ritual: Essays on Face-to-Face Behavior*. New York: Pantheon, 1967.

Goffman, Erving. "On Face-Work: An Analysis of Ritual Elements in Social Interaction." *Psychiatry* 18 (1955): 213–231.

Goffman, Erving. *The Presentation of the Self in Everyday Life*. New York: Doubleday, 1959.

Gomez, Gay M. "Perspective, Power, and Priorities: New Orleans and the Mississippi River Flood of 1927." In *Transforming New Orleans and Its Environs: Centuries of Change*, edited by Craig E. Colten, 109–120. Pittsburgh: University of Pittsburgh Press, 2000.

Graham, Stephen, and Simon Marvin. *Splintering Urbanism: Networked Infrastructures, Technological Mobilities and the Urban Condition*. London: Routledge, 2001.

Graham, Stephen, and Nigel Thrift. "Out of Order: Understanding Repair and Maintenance." *Theory, Culture and Society* 24, no. 3 (2007): 1–25.

Granberry, Michael. "20 Years Later, Span Is Loved and Loathed." *Los Angeles Times*, August 2, 1989. http://articles.latimes.com/1989-08-02/local/me-406_1_toll-bridge.

Gray Plant Mooty. "Investigative Report to Joint Committee to Investigate the I-35W Bridge Collapse." May 2008. https://www.leg.state.mn.us/docs/2008/other/080513/Investigative_Report.pdf.

Greenfield, Adam. "A Sociology of the Smartphone." *Longreads* (blog), June 13, 2017. https://longreads.com/2017/06/13/a-sociology-of-the-smartphone/.

Griswold del Castillo, Richard. *The Los Angeles Barrio, 1850–1890: A Social History*. Berkeley: University of California Press, 1979.

Hannigan, John. *Environmental Sociology: A Social Constructionist Perspective*. 2nd ed. New York: Routledge, 2006.

Haraway, Donna J. *Simians, Cyborgs, and Women: The Reinvention of Nature*. New York: Routledge, 1991.

Haraway, Donna J. *Staying with the Trouble: Making Kin in the Chthulucene*. Chapel Hill, NC: Duke University Press, 2016.

Harper, Douglas. *Working Knowledge: Skill and Community in a Small Shop*. Berkeley: University of California Press, 1987.

Harvey, David. *A Brief History of Neoliberalism*. Oxford: Oxford University Press, 2007.

Harvey, David. *The Condition of Postmodernity: An Enquiry into the Origins of Cultural Change*. Cambridge, MA: Blackwell, 1990.

Harvey, Penny, and Hannah Knox. *Roads: An Anthropology of Infrastructure and Expertise*. Ithaca, NY: Cornell University Press, 2015.

Henke, Christopher R. *Cultivating Science, Harvesting Power: Science and Industry in California Agriculture*. Cambridge, MA: MIT Press, 2008.

Henke, Christopher R. "The Mechanics of Workplace Order: Toward a Sociology of Repair." *Berkeley Journal of Sociology* 44 (2000): 55–81.

Henke, Christopher R. "Situation Normal? Repairing a Risky Ecology." *Social Studies of Science* 37, no. 1 (2007): 135–142.

Henke, Christopher R. "The Sustainable University: Repair as Maintenance and Transformation." *Continent* 6, no. 1 (2017): 40–45.

Henke, Christopher R., and Thomas F. Gieryn. "Sites of Scientific Practice: The Enduring Importance of Place." In *New Handbook of Science and Technology Studies*, edited by Edward Hackett, Olga Amsterdamska, Michael Lynch, and Judy Wajcman. Cambridge, MA: MIT Press, 2007.

Herrera, Juan Felipe. "Logan Heights and the World." In *100 Great Poems: Classic Poets and Beatnik Freaks*, track 6. Words of Wisdom Records, 2012.

Highsmith, Andrew R. *Demolition Means Progress: Flint, Michigan, and the Fate of the American Metropolis*. Chicago: University of Chicago Press, 2015.

Hochschild, Arlie Russell. *The Managed Heart: The Commercialization of Human Feeling*. Berkeley: University of California Press, 1983.

Hoof, J. van. "Forty Years of Fanger's Model of Thermal Comfort: Comfort for All?" *Indoor Air* 18, no. 3 (2008): 182–201.

Houston, Lara. "Mobile Phone Repair Knowledge in Downtown Kampala: Local and Trans-Local Circulations." In *Repair Work Ethnographies: Revisiting Breakdown, Relocating Materiality*, edited by Ignaz Strebel, Alain Bovet, and Philippe Sormani, 129–160. Singapore: Palgrave Macmillan, 2019.

Houston, Lara. "The Timeliness of Repair." *Continent* 6, no. 1 (2017): 51–55.

Houston, Lara. "Unsettled Repair Tools: The 'Death' of the J.A.F. Box." Paper presented at the Maintainers Conference, Stevens Institute of Technology, Hoboken, NJ, 2016, 1–13. https://static1.squarespace.com/static/56a8e2fca12f446482d67a7a/t/570e9c8f01dbae9c3322fa7d/1460575376524/Maintainers-Lara-Houston.pdf.

Houston, Lara, and Steven J. Jackson. "Caring for the 'Next Billion' Mobile Handsets: Opening Proprietary Closures Through the Work of Repair." In *Proceedings of the Eighth International Conference on Information and Communication Technologies and Development*, 10:1–10:11. New York: ACM, 2016.

Houston, Lara, Steven J. Jackson, Daniela K. Rosner, Syed Ishtiaque Ahmed, Meg Young, and Laewoo Kang. "Values in Repair." In *Proceedings of the 2016 CHI Conference on Human Factors in Computing Systems*, 1403–1414. New York: ACM Press, 2016.

Houston, Lara, Daniela K. Rosner, Steven J. Jackson, and Jamie Allen, eds. "R3pair Volume." Special issue, *Continent*, no. 6.1 (2017). http://continentcontinent.cc/index.php/continent/issue/view/27.

Hughes, Everett C. "Work and the Self." In *Social Psychology at the Crossroads*, edited by John H. Rohrer and Muzafer Sherif, 313–323. New York: Harper and Brothers, 1951.

Hughes, Thomas P. *Networks of Power: Electrification in Western Society, 1880–1930*. Baltimore: Johns Hopkins University Press, 1983.

Hughes, Thomas P. *Rescuing Prometheus: Four Monumental Projects That Changed the Modern World*. New York: Pantheon Books, 1998.

International Commission on Stratigraphy. "ICS—GSSPs." Accessed December 13, 2018, at http://www.stratigraphy.org/index.php/ics-gssps.

Irani, Lilly. "The Cultural Work of Microwork." *New Media and Society* 17, no. 5 (2015): 720–739.

Irani, Lilly. "Microworking the Crowd." *Limn*, February 13, 2012, https://limn.it/articles/microworking-the-crowd/.

Jackson, Steven J. "Rethinking Repair." In *Media Technologies: Essays on Communication, Materiality, and Society*, edited by Tarleton Gillespie, Pablo Boczkowski, and Kirsten Foot, 221–239. Cambridge, MA: MIT Press, 2014.

Jackson, Steven J., Paul N. Edwards, Geoffrey C. Bowker, and Cory P. Knobel. "Understanding Infrastructure: History, Heuristics and Cyberinfrastructure Policy." *First Monday* 12, no. 6 (2007). http://www.firstmonday.dk/ojs/index.php/fm/article/view/1904.

Jackson, Steven J., Alex Pompe, and Gabriel Krieshok. "Repair Worlds: Maintenance, Repair, and ICT for Development in Rural Namibia." In *Proceedings of the ACM 2012 Conference on Computer Supported Cooperative Work*, 107. New York: ACM Press, 2012.

Jones, Rhys, and Peter Merriman. "Hot, Banal and Everyday Nationalism: Bilingual Road Signs in Wales." *Political Geography* 28, no. 3 (2009): 164–173.

Kallis, Giorgos. *Degrowth*. Newcastle upon Tyne: Agenda Publishing, 2018.

Karjalainen, S. "Thermal Comfort and Gender: A Literature Review." *Indoor Air* 22, no. 2 (2012): 96–109.

Kates, Robert W., Thomas M. Parris, and Anthony A. Leiserowitz. "What Is Sustainable Development? Goals, Indicators, Values, and Practice." *Environment* 47, no. 3 (2005): 8–21.

Kelman, Ari. *A River and Its City: The Nature of Landscape in New Orleans*. Berkeley: University of California Press, 2003.

Kingma, Boris, and Wouter van Marken Lichtenbelt. "Energy Consumption in Buildings and Female Thermal Demand." *Nature Climate Change* 5 (2015): 1054–1056.

Kingsworth, Paul, and Dougald Hine. "The Manifesto." Dark Mountain, 2009. https://dark-mountain.net/about/manifesto/.

Koebler, Jason. "Why American Farmers Are Hacking Their Tractors with Ukrainian Firmware." *Motherboard* (blog), March 21, 2017. https://motherboard.vice.com/en_us/article/xykkkd/why-american-farmers-are-hacking-their-tractors-with-ukrainian-firmware.

Kuchinskaya, Olga. *The Politics of Invisibility: Public Knowledge about Radiation Health Effects after Chernobyl*. Cambridge, MA: MIT Press, 2014.

Laet, Marianne de, and Annemarie Mol. "The Zimbabwe Bush Pump: Mechanics of a Fluid Technology." *Social Studies of Science* 30, no. 2 (2000): 225–263.

Lampland, Martha, and Susan Leigh Star, eds. *Standards and Their Stories: How Quantifying, Classifying and Formalizing Practices Shape Everyday Life*. Ithaca, NY: Cornell University Press, 2008.

Larkin, Brian. "The Politics and Poetics of Infrastructure." *Annual Review of Anthropology* 42, no. 1 (2013): 327–343.

Latour, Bruno. "Give Me a Laboratory and I Will Raise the World." In *Science Observed*, edited by K. Knorr and M. Mulkay, 141–170. Beverly Hills: Sage, 1983.

Latour, Bruno. *Pandora's Hope: Essays on the Reality of Science Studies*. Cambridge, MA: Harvard University Press, 1999.

Latour, Bruno. *The Pasteurization of France*. Cambridge, MA: Harvard University Press, 1984.

Latour, Bruno. *Reassembling the Social: An Introduction to Actor-Network-Theory*. Oxford: Oxford University Press, 2005.

Latour, Bruno. *Science in Action: How to Follow Scientists and Engineers through Society*. Cambridge, MA: Harvard University Press, 1987.

Latour, Bruno. *We Have Never Been Modern*. Cambridge, MA: Harvard University Press, 1993.

Latour, Bruno. "Where Are the Missing Masses? The Sociology of a Few Mundane Artifacts." In *Shaping Technology/Building Society*, edited by Wiebe E. Bijker and John Law, 225–258. Cambridge, MA: MIT Press, 1992.

Law, John. *Aircraft Stories: Decentering the Object in Technoscience*. Durham, NC: Duke University Press, 2002.

Law, John. "The Materials of STS." In *The Oxford Handbook of Material Culture Studies*, edited by Dan Hicks and Mary C. Beaudry, 173–188. New York: Oxford University Press, 2010.

Law, John. *Organizing Modernity*. Cambridge, MA: Blackwell, 1994.

Law, John. "Technology and Heterogeneous Engineering: The Case of Portuguese Expansion." In *The Social Construction of Technological Systems: New Directions in the Sociology and History of Technology*, edited by Wiebe E. Bijker, Thomas P. Hughes, and Trevor J. Pinch, 111–134. Cambridge, MA: MIT Press, 1987.

Leibovich, Mark. "Armageddon Moves Inside the Computer: Los Alamos Is Calculating a New Nuclear Testing Era." *Washington Post*, November 18, 1998.

Lemke, Thomas. "New Materialisms: Foucault and the 'Government of Things.'" *Theory, Culture and Society* 32, no. 4 (2015): 3–25.

Lengnick, Laura. *Resilient Agriculture: Cultivating Food Systems for a Changing Climate.* Philadelphia: New Society Publishers, 2015.

LePatner, Barry B. *Too Big to Fall: America's Failing Infrastructure and the Way Forward.* New York: Foster Publishing, 2010.

Leslie, Stuart, and Indira Chowdhury. "Homi Bhabha, Master Builder of Nuclear India." *Physics Today* 71, no. 9 (2018): 48.

Levy, Paul. "4 Dead, 79 Injured, 20 Missing after Dozens of Vehicles Plummet into River." *Minneapolis Star-Tribune*, August 2, 2007. http://www.startribune.com/4-dead -79-injured-20-missing-after-dozens-of-vehicles-plummet-into-river/11593606.

Lewis, Peirce F. *New Orleans: The Making of an Urban Landscape.* 2nd ed. Santa Fe, NM: Center for American Places, 2003.

Lewis, Simon L., and Mark A. Maslin. *The Human Planet: How We Created the Anthropocene.* New Haven, CT: Yale University Press, 2018.

Librová, Hana, and Vojtěch Pelikán. "Ethical Motivations and the Phenomenon of Disappointment in Two Types of Environmental Movements: Neo-Environmentalism and the Dark Mountain Project." *Environmental Values* 25, no. 2 (2016): 167–193.

Lowder, Sarah K., Jakob Skoet, and Saumya Singh. "What Do We Really Know about the Number and Distribution of Farms and Family Farms Worldwide? Background Paper for *The State of Food and Agriculture 2014*." Rome: United Nations, Food and Agriculture Organization, 2014.

Lynch, Michael. "Against Reflexivity as an Academic Virtue and Source of Privileged Knowledge." *Theory, Culture and Society* 17, no. 3 (2000): 26–54.

Mann, Michael. "The Autonomous Power of the State: Its Origins, Mechanisms and Results." *European Journal of Sociology* 25, no. 2 (1984): 185–213.

Mann, Michael. *The Sources of Social Power*, vol. 2: *The Rise of Classes and Nation States, 1760–1914.* Cambridge: Cambridge University Press, 1993.

Marcus, Alan I. *Agricultural Science and the Quest for Legitimacy: Farmers, Agricultural Colleges, and Experiment Stations, 1870–1890.* Ames: Iowa State University Press, 1985.

Martin, Judith A. "Neighborhoods Confront a Disaster Aftermath." In *The City, the River, the Bridge: Before and After the Minneapolis Bridge Collapse*, edited by Patrick Nunnally, 57–75. Minneapolis: University of Minnesota Press, 2011.

Masco, Joseph. *The Nuclear Borderlands: The Manhattan Project in Post–Cold War New Mexico*. Princeton, NJ: Princeton University Press, 2006.

McCourt, David M. *Britain and World Power since 1945: Constructing a Nation's Role in International Politics*. Ann Arbor: University of Michigan Press, 2015.

Merrill, Dave, and Lauren Leatherby. "Here's How America Uses Its Land." *Bloomberg*, July 31, 2018. https://www.bloomberg.com/graphics/2018-us-land-use/.

Merriman, Peter, and Rhys Jones. "Nations, Materialities and Affects." *Progress in Human Geography* 41, no. 5 (2017): 600–617.

Mitchell, Timothy. *Carbon Democracy: Political Power in the Age of Oil*. London: Verso Books, 2011.

Mohl, Raymond A. "The Interstates and the Cities: The U.S. Department of Transportation and the Freeway Revolt, 1966–1973." *Journal of Policy History* 20, no. 2 (2008): 193–226.

Mohl, Raymond A. "Stop the Road: Freeway Revolts in American Cities." *Journal of Urban History* 30, no. 5 (2004): 674–706.

Mol, Annemarie. *The Logic of Care: Health and the Problem of Patient Choice*. New York: Routledge, 2008.

Mol, Annemarie, Ingunn Moser, and Jeanette Pols. "Care: Putting Practice into Theory." In *Care in Practice: On Tinkering in Clinics, Homes, and Farms*, edited by Annemarie Mol, Ingunn Moser, and Jeanette Pols, 7–26. Bielefeld, Germany: Transcript Verlag, 2010.

Moore, Kelly. *Disrupting Science: Social Movements, American Scientists, and the Politics of the Military, 1945–1975*. Princeton, NJ: Princeton University Press, 2008.

Mukerji, Chandra. *A Fragile Power: Scientists and the State*. Princeton, NJ: Princeton University Press, 1989.

Mukerji, Chandra. *Impossible Engineering: Technology and Territoriality on the Canal du Midi*. Princeton, NJ: Princeton University Press, 2009.

Mukerji, Chandra. "Intelligent Uses of Engineering and the Legitimacy of State Power." *Technology and Culture* 44, no. 4 (2003): 655–676.

Mukerji, Chandra. "The Landscape Garden as Material Culture: Lessons from France." In *The Oxford Handbook of Material Culture Studies*, edited by Dan Hicks and Mary C. Beaudry, 543–561. New York: Oxford University Press, 2010.

Mukerji, Chandra. *Modernity Reimagined: An Analytic Guide*. New York: Routledge, 2017.

Mukerji, Chandra. *Territorial Ambitions and the Gardens of Versailles*. Cambridge: Cambridge University Press, 1997.

Nestle, Marion. *Food Politics: How the Food Industry Influences Nutrition and Health.* Berkeley: University of California Press, 2002.

Nickerson, Cynthia, and Allison Borchers. "How Is Land in the United States Used? A Focus on Agricultural Land." Washington, DC: Economic Research Service, US Department of Agriculture, 2012. https://www.ers.usda.gov/amber-waves/2012/march/data-feature-how-is-land-used/.

Norman, Don. *The Design of Everyday Things: Revised and Expanded Edition.* New York: Basic Books, 2013.

Norris, Frank. "Logan Heights: Growth and Change in the Old 'East End.'" *Journal of San Diego History* 29, no. 1 (1983). http://sandiegohistory.org/journal/1983/january/logan/.

Nunnally, Patrick. "Building the New Bridge: Process and Politics in City-Building." In *The City, the River, the Bridge: Before and After the Minneapolis Bridge Collapse,* edited by Patrick Nunnally, 35–54. Minneapolis: University of Minnesota Press, 2011.

Orr, Julian. *Talking about Machines: An Ethnography of a Modern Job.* Ithaca, NY: Cornell University Press, 1996.

Penny, Dan, Cameron Zachreson, Roland Fletcher, David Lau, Joseph T. Lizier, Nicholas Fischer, Damian Evans, et al. "The Demise of Angkor: Systemic Vulnerability of Urban Infrastructure to Climatic Variations." *Science Advances* 4, no. 10 (2018): eau4029.

Perez, Caroline Criado. *Invisible Women: Data Bias in a World Designed for Men.* New York: Abrams Press, 2019.

Perrow, Charles. *Normal Accidents: Living with High-Risk Technologies.* 2nd ed. Princeton, NJ: Princeton University Press, 1999.

Petroski, Henry. *The Road Taken: The History and Future of America's Infrastructure.* New York: Bloomsbury, 2016.

Pickett, Brent L. "Foucault and the Politics of Resistance." *Polity* 28, no. 4 (1996): 445–466.

Pires, Larissa. "Gender in the Modernist City: Shaping Power Relations and National Identity with the Construction of Brasília." PhD diss., Iowa State University, 2013.

Pollan, Michael. *The Omnivore's Dilemma: A Natural History of Four Meals.* New York: Penguin, 2006.

Quito, Anne. "The New Fountain at Versailles Was Inspired by 17th-Century Power Games." *Quartz,* 2015. https://qz.com/511706/the-new-fountain-at-versailles-was-inspired-by-17th-century-power-games/.

Reilly, J., F. Tubiello, B. McCarl, D. Abler, R. Darwin, K. Fuglie, S. Hollinger, et al. "U.S. Agriculture and Climate Change: New Results." *Climatic Change* 57, no. 1/2 (2003): 43–67.

"Repair, n.2." In *OED Online*. Oxford: Oxford University Press, December 2018. http://www.oed.com/view/Entry/162629.

Ribes, David, and Thomas A. Finholt. "The Long Now of Technology Infrastructure: Articulating Tensions in Development." *Journal of the Association for Information Systems* 10, no. 5 (2009): 375–398.

Rice, Dave. "Barrio Logan Group Says Basta to Stadium Rhetoric." *San Diego Reader*, July 15, 2016. https://www.sandiegoreader.com/news/2016/jul/15/ticker-community -fighters-calls-bull-t-stadium/.

"Right to Repair—iFixit." Accessed December 3, 2018, at https://ifixit.org/right.

Rosen, Martin D., and James Fisher. "Chicano Park and the Chicano Park Murals: Barrio Logan, City of San Diego, California." *Public Historian* 23, no. 4 (2001): 91–112.

Rosenberg, Charles E. "Rationalization and Reality in the Shaping of American Agricultural Research, 1875–1914." *Social Studies of Science* 7 (1977): 401–422.

Rosner, Daniela, and Jonathan Bean. "Learning from IKEA Hacking: I'm Not One to Decoupage a Tabletop and Call It a Day." In *Proceedings of the SIGCHI Conference on Human Factors in Computing Systems*, 419–422. New York: ACM, 2009.

Rosner, Daniela K., and Morgan Ames. "Designing for Repair? Infrastructures and Materialities of Breakdown." In *Proceedings of the 17th ACM Conference on Computer Supported Cooperative Work and Social Computing*, 319–331. New York: ACM, 2014.

Ross, Andrew. "Universities and the Urban Growth Machine." *Dissent Magazine*, October 4, 2012. http://www.dissentmagazine.org/online_articles/universities-and -the-urban-growth-machine.

Ruder, Cynthia A. *Making History for Stalin: The Story of the Belomor Canal*. Gainesville: University Press of Florida, 1998.

Russell, Andrew L. *Open Standards and the Digital Age: History, Ideology, and Networks*. Cambridge: Cambridge University Press, 2014.

Russell, Andrew L., and Lee Vinsel. "After Innovation, Turn to Maintenance." *Technology and Culture* 59, no. 1 (2018): 1–25.

Russell, Andrew L., and Lee Vinsel. "Hail the Maintainers." *Aeon*, 2016. https://aeon.co/ essays/innovation-is-overvalued-maintenance-often-matters-more.

Russell, Andrew L., and Lee Vinsel. "Let's Get Excited about Maintenance!" *New York Times*, July 22, 2017.

Saulny, Susan, and Jennifer Steinhauer. "Bridge Collapse Revives Issue of Road Spending." *New York Times*, August 7, 2007. http://www.nytimes.com/2007/08/07/us/07highway.html.

Sayes, Edwin. "Actor–Network Theory and Methodology: Just What Does It Mean to Say That Nonhumans Have Agency?" *Social Studies of Science* 44, no. 1 (2014): 134–149.

Schegloff, Emanuel A. "Repair after Next Turn: The Last Structurally Provided Defense of Intersubjectivity in Conversation." *American Journal of Sociology* 97, no. 5 (1992): 1295–1345.

Schegloff, Emanuel A. "Third Turn Repair." In *Towards a Social Science of Language: Papers in Honor of William Labov*, edited by G. R. Guy, M. C. Feagin, D. Schiffrin, and J. Baugh, 31–40. Amsterdam: John Benjamins, 1997.

Schegloff, Emanuel A., Gail Jefferson, and Harvey Sacks. "The Preference for Self-Correction in the Organization of Repair for Conversation." *Language* 53, no. 2 (1977): 361–382.

Schlosser, Eric. *Command and Control: Nuclear Weapons, the Damascus Accident, and the Illusion of Safety*. Reprint ed. New York: Penguin Books, 2014.

Schwartz, Stephen I. *Atomic Audit: The Costs and Consequences of U.S. Nuclear Weapons since 1940*. Washington, DC: Brookings Institution Press, 1998.

Scott, James C. *The Moral Economy of the Peasant: Rebellion and Subsistence in Southeast Asia*. New Haven: Yale University Press, 1976.

Scott, James C. *Seeing Like a State: How Certain Schemes to Improve the Human Condition Have Failed*. New Haven: Yale University Press, 1998.

Scott, James C. *Weapons of the Weak: Everyday Forms of Peasant Resistance*. New Haven: Yale University Press, 1987.

Shapin, Steven. "The Invisible Technician." *American Scientist* 77 (1989): 554–563.

Shapin, Steven. *A Social History of Truth: Civility and Science in Seventeenth-Century England*. Chicago: University of Chicago Press, 1994.

Shapin, Steven, and Simon Schaffer. *Leviathan and the Air-Pump: Hobbes, Boyle, and the Experimental Life*. Princeton, NJ: Princeton University Press, 1985.

Shever, Elana. *Resources for Reform: Oil and Neoliberalism in Argentina*. Palo Alto, CA: Stanford University Press, 2012.

Sims, Benjamin. "Concrete Practices: Testing in an Earthquake-Engineering Laboratory." *Social Studies of Science* 29, no. 4 (1999): 483–518.

Sims, Benjamin. "'The Day after the Hurricane': Infrastructure, Order, and the New Orleans Police Department's Response to Hurricane Katrina." *Social Studies of Science* 37, no. 1 (2007): 111–118.

Sims, Benjamin. "Disoriented City: Infrastructure, Social Order, and the Police Response to Hurricane Katrina." In *Disrupted Cities: When Infrastructure Fails*, edited by Stephen Graham, 41–53. Milton Park, UK: Routledge, 2010.

Sims, Benjamin. "Layers of Abstraction and the Organization of Repair in High Performance Computing." Paper presented at the Society for Social Studies of Science Annual Meeting, New Orleans, LA, 2019.

Sims, Benjamin. "Making Technological Timelines: Anticipatory Repair and Testing in High Performance Scientific Computing." *Continent* 6, no. 1 (2017): 81–84.

Sims, Benjamin. "On Shifting Ground: Earthquakes, Retrofit and Engineering Culture in California." PhD diss., University of California, San Diego, 2000.

Sims, Benjamin. "Safe Science: Material and Social Order in Laboratory Work." *Social Studies of Science* 35, no. 3 (2005): 333–366.

Sims, Benjamin. "Seismic Shifts and Retrofits: Scale and Complexity in the Seismic Retrofit of California Bridges." In *Retrofitting Cities: Priorities, Governance and Experimentation*, edited by Mike Hodson and Simon Marvin, 13–33. London: Routledge, 2016.

Sims, Benjamin. "Things Fall Apart: Disaster, Infrastructure, and Risk." *Social Studies of Science* 37, no. 1 (2007): 93–95.

Sims, Benjamin, and Christopher R. Henke. "Maintenance and Transformation in the U.S. Nuclear Weapons Complex." *IEEE Technology and Society Magazine* 27, no. 3 (2008): 32–38.

Sims, Benjamin, and Christopher R. Henke. "Repairing Credibility: Repositioning Nuclear Weapons Knowledge after the Cold War." *Social Studies of Science* 42, no. 3 (2012): 324–347.

Smith, Joshua Emerson. "San Diego Redrafting Highly Contested Blueprint for Barrio Logan." *San Diego Union-Tribune*, August 28, 2017. https://www.sandiegouniontribune.com/news/environment/sd-me-barrio-logan-update-20170825-story.html.

Spielmaker, D. M. "Growing a Nation: The Story of American Agriculture." 2018. https://www.agclassroom.org/gan/timeline/index.htm.

Sportelli, Natalie. "Colgate University's Trudy Fitness Center Awarded Gold LEED Certification." *Colgate University News* (blog), June 27, 2012. http://news.colgate.edu/2012/06/trudy-fitness-center-awarded-gold-leed-certification.html/.

Stalmer, Julie. "Barrio Logan Community Plan Not Updated Much." *San Diego Reader*. September 23, 2017. https://www.sandiegoreader.com/news/2017/sep/22/stringers-barrio-logan-community-plan-not-updated/.

"Stand Up for Your Right to Repair." Repair Association. Accessed December 3, 2018, at http://repair.org/stand-up/.

Star, Susan Leigh. "The Ethnography of Infrastructure." *American Behavioral Scientist* 43, no. 3 (1999): 377–391.

Star, Susan Leigh, and Karen Ruhleder. "Steps toward an Ecology of Infrastructure: Design and Access for Large Information Spaces." *Information Systems Research* 7, no. 1 (1996): 111–134.

Steffen, Will, Wendy Broadgate, Lisa Deutsch, Owen Gaffney, and Cornelia Ludwig. "The Trajectory of the Anthropocene: The Great Acceleration." *Anthropocene Review* 2, no. 1 (2015): 81–98.

Steinhardt, Stephanie B., and Steven J. Jackson. "Anticipation Work: Cultivating Vision in Collective Practice." In *Proceedings of the 18th ACM Conference on Computer Supported Cooperative Work and Social Computing*, 443–453. New York: ACM, 2015.

Strebel, Ignaz, Alain Bovet, and Philippe Sormani, eds. *Repair Work Ethnographies: Revisiting Breakdown, Relocating Materiality*. Singapore: Palgrave Macmillan, 2019.

Summerson Carr, E., and Michael Lempert. *Scale: Discourse and Dimensions of Social Life*. Berkeley: University of California Press, 2016.

Tanweer, Anissa, Brittany Fiore-Gartland, and Cecilia Aragon. "Impediment to Insight to Innovation: Understanding Data Assemblages through the Breakdown-Repair Process." *Information, Communication and Society* 19, no. 6 (2016): 736–752.

Thomas, W. I. *The Child in America*. New York: Knopf, 1928.

Thomas, W. I. *The Unadjusted Girl*. Boston: Little, Brown, 1923.

Turner, Fred. *From Counterculture to Cyberculture: Stewart Brand, the Whole Earth Network, and the Rise of Digital Utopianism*. Chicago: University of Chicago Press, 2006.

Ureta, Sebastián. *Assembling Policy: Transantiago, Human Devices, and the Dream of a World-Class Society*. Cambridge, MA: MIT Press, 2015.

Ureta, Sebastián. "Normalizing Transantiago: On the Challenges (and Limits) of Repairing Infrastructures." *Social Studies of Science* 44, no. 3 (2014): 368–392.

Ureta, Sebastián. "Waiting for the Barbarians: Disciplinary Devices on Metro de Santiago." *Organization* 20, no. 4 (2013): 596–614.

US Department of Agriculture, Economic Research Service. "Ag and Food Sectors and the Economy," 2018. https://www.ers.usda.gov/data-products/ag-and-food-statistics -charting-the-essentials/ag-and-food-sectors-and-the-economy.aspx.

US Department of Agriculture, National Agricultural Statistics Service. "USDA Census of Agriculture Historical Archive," 2018. http://agcensus.mannlib.cornell.edu/ AgCensus/homepage.do;jsessionid=5BFA0828913934BE4F1B693340F0AC34.

US Environmental Protection Agency. "National Summary of State Information: Water Quality Assessment and TMDL Information," 2018. https://ofmpub.epa.gov/waters10/attains_nation_cy.control#prob_source.

US Environmental Protection Agency. "Sources of Greenhouse Gas Emissions." Overviews and Factsheets. US EPA, December 29, 2015. https://www.epa.gov/ghgemissions/sources-greenhouse-gas-emissions.

US Global Change Research Program. "Fourth National Climate Assessment." Washington, DC, 2018. https://nca2018.globalchange.gov.

US Green Building Council. "LEED | U.S. Green Building Council," 2016. http://www.usgbc.org/leed.

US National Transportation Safety Board. "Collapse of the I-35W Highway Bridge, Minneapolis, Minnesota, August 1, 2007." November 14, 2008.

Villa, Raúl Homero. *Barrio Logos: Space and Place in Urban Chicano Literature and Culture.* Austin: University of Texas Press, 2000.

von Schnitzler, Antina. *Democracy's Infrastructure: Techno-Politics and Protest after Apartheid.* Princeton, NJ: Princeton University Press, 2016.

von Schnitzler, Antina. "Infrastructure, Apartheid Technopolitics, and Temporalities of 'Transition.'" In *The Promise of Infrastructure*, edited by Nikhil Anand, Akhil Gupta, and Hannah Appel, 133–154. Durham, NC: Duke University Press, 2018.

Wallace, Bill, and Michael Taylor. "State Was Slow to Reinforce the I-880 Pillars, Experts Say." *San Francisco Chronicle*, October 20, 1989.

Walthall, C. L., J. Hatfield, P. Backlund, and L. Lengnick. "Climate Change and Agriculture in the United States: Effects and Adaptation." Washington, DC: USDA, 2012. https://www.usda.gov/oce/climate_change/effects_2012/CC%20and%20Agriculture%20Report%20(02-04-2013)b.pdf.

Warth, Gary. "Chicano Park Named National Historic Landmark." *San Diego Union-Tribune*, January 11, 2017. https://www.sandiegouniontribune.com/news/politics/sd-me-chicano-historic-20170111-story.html.

Waters, Colin N., Jan Zalasiewicz, Colin Summerhayes, Anthony D. Barnosky, Clément Poirier, Agnieszka Gałuszka, Alejandro Cearreta, et al. "The Anthropocene Is Functionally and Stratigraphically Distinct from the Holocene." *Science* 351, no. 6269 (2016): 137.

Weber, Max. "Bureaucracy." In *From Max Weber: Essays in Sociology*, edited by H. H. Gerth and C. Wright Mills, 196–244. New York: Oxford University Press, 1946.

Weber, Max. *The Protestant Ethic and the Spirit of Capitalism.* New York: Routledge, 1930.

Wellerstein, Alex. "Maintaining the Bomb." *Restricted Data: The Nuclear Secrecy Blog* (blog), April 8, 2016, http://blog.nuclearsecrecy.com/2016/04/08/maintaining-the -bomb/.

Wharton, Amy S. "The Sociology of Emotional Labor." *Annual Review of Sociology* 35 (2009): 147–65.

White, Richard. *The Organic Machine.* New York: Hill and Wang, 1995.

Winner, Langdon. "Do Artifacts Have Politics?" In *The Social Shaping of Technology*, edited by Donald MacKenzie and Judy Wajcman, 26–38. Milton Keynes, UK: Open University Press, 1985.

World Commission on Environment and Development. *Our Common Future.* New York: Oxford University Press, 1987.

Wynne, Brian. "May the Sheep Safely Graze? A Reflexive View of the Expert-Lay Divide." In *Risk, Environment, and Modernity: Towards a New Ecology*, edited by Scott Lash, Bronislaw Szerszynski, and Brian Wynne, 44–83. Thousand Oaks, CA: Sage, 1996.

Wynne, Brian. "Sheepfarming after Chernobyl: A Case Study in Communicating Scientific Information." *Environment* 31 (1989): 10–15, 33–39.

Yang, Liu, Haiyan Yan, and Joseph C. Lam. "Thermal Comfort and Building Energy Consumption Implications—a Review." *Applied Energy* 115 (2014): 164–173.

Zalasiewicz, Jan, Colin N. Waters, Mark Williams, Anthony D. Barnosky, Alejandro Cearreta, Paul Crutzen, Erle Ellis, et al. "When Did the Anthropocene Begin? A Mid-Twentieth Century Boundary Level Is Stratigraphically Optimal." *Quaternary International* 383 (2015): 196–203.

Index